SOME TRUST IN CHARIOTS

DO YOU BELIEVE THAT GOD WAS AN ASTRONAUT FROM OUTER SPACE?

Chariots of the Gods has set off a revolution in our thinking—a vision of our distant past so violently different from what we have previously imagined that it has left the minds of millions stunned and confused, torn between doubt and belief.

Now into this gigantic controversy have stepped sixteen authorities with impeccable credentials and total objectivity.

Here at last, from sources that cannot be denied, are the authentic facts that will allow you to confidently make up your mind about *Chariots of the Gods* and all that it means to us today.

SOME TRUST IN CHARIOTS

EDITED BY BARRY THIERING
AND EDGAR CASTLE

Sixteen views on Erich von Däniken's
Chariots of the Gods?

POPULAR LIBRARY • NEW YORK

All POPULAR LIBRARY books are carefully selected by the POPULAR LIBRARY Editorial Board and represent titles by the world's greatest authors.

POPULAR LIBRARY EDITION

Published by arrangement with Bailey Brothers and Swinfen Ltd.

PRINTED IN THE UNITED STATES OF AMERICA

Contents

SOME TRUST IN CHARIOTS

Preface

Chariots of the Gods? by the Swiss Erich von Däniken is a book of popular archaeology, sensational and rhetorical in style, which has had an extraordinarily wide following since its publication in 1967. The writers are varied in their assumptions about the nature of things: among them there are Christians of several kinds, a member of the Jewish community, and at least one agnostic. But they are closely united in believing that truth is hard-won and precious, and that *Chariots of the Gods?* requires a counter-attack. The editors are deeply grateful to the contributors for their speedy and generous response to what was virtually a call to arms. For artwork, thanks are offered to Robert Graham, T. M. R. Castle, and the Art Department of Griffin Press, Page references throughout this book are to the Corgi paper-back edition of *Chariots of the Gods?* (London, 1971).

Introduction: The Case for the Prosecution

THE EDITORS

Edgar Castle, B.A. (Adelaide), B.Litt (New England), Dip.Ed. (Durham), M.A.C.E., has been a schoolmaster for about twenty years. He has published a couple of school books, writes for periodical publication, and is active in several professional organisations.

The Revd B. B. Thiering, M.A. (Sydney), Th. L., Dip. R.E. (Melbourne), M.A.C.E., Chaplain and Head of General Studies, Cranbrook School, is also Secretary to the N.S.W. Committee for Religious Education in Independent Schools. He was previously Rector of St Stephen's Church, Mittagong.

Introduction: The Case for the Prosecution

> "The author's theory is that in the earth's remote past the planet was visited by beings from space who perhaps fathered humanity as we know it. A challenging contribution to discussion about our past—and future." . . . "Powerful stuff which, no matter how one tries, cannot be discarded as crack-pot theories."

A book of which these claims were true should be read, and urgently read, by even more people than the tens of thousands who have already read Erich von Däniken's *Chariots of the Gods?* For now that the human race is treading the brink of total disaster by ecological upset or by nuclear war, a book which can point to the probability of extra-terrestrial aid in the past may also imply the hope of extra-terrestrial aid in the terrifying future. If, that is to say, the reviewers whose praise ornaments the cover of the popular edition of the book are *right*, Erich von Däniken is suggesting a programme that might save us from everything we have to fear.

If, however, the reviewers are *wrong*, then the book is at best folly, and at worst a cruel delusion.

If the "power" of von Däniken's theory is only the power of audacity and rhetoric, then it does the human race a great disservice by breeding false optimism and distracting it from real and urgent scientific and moral tasks. A cargo cult in a primative country is pathetic: a cargo cult in the developed half of the world is potential suicide.

This present book has been assembled in the belief that the reviewers are *wrong*. It was conceived after its editors had followed fashion and read von Däniken. They were shocked by the looseness of the arguments, and amazed by the entourage the book had gathered. Above all, they were saddened that a very mildly expressed questioning of the book in the press drew vehement responses.

"Shocked", "amazed", and "saddened" are strong words but true. For the success of *Chariots of the Gods?* after a century of free and compulsory state education in most

Western countries *is* amazing, shocking and saddening.

It seemed that a thorough, definitive and public examination of von Däniken's shadowy thesis ought to be made as quickly as possible. Two booklets had already been issued against von Däniken in Australia; but these, though damaging to him, were relatively slight and were open to the charge that they were written from particular religious viewpoints. It was also suggested that "many people will continue to think that von Däniken is saying something worthwhile" whatever was said or written. But there seemed to be at least a case for trying to reduce the number.

A few enquiries showed the editors that they were not alone in fearing a *Chariots of the Gods?* cult. Indeed, the willingness of some very busy experts to lend their knowledge and their skill to answering a book which they might have thought beneath their notice, was heartening. The editors found themselves in the humbling position of having to marshal the evidence of several scholars, any one of whom was much more clever than they.

Inevitably, there must be some editorial imperfections in this book which resulted. Since the contributors were asked to write freely about whatever they thought relevant there is some duplication of information, and there are some few issues raised by von Däniken which escape examination. The quantity of matter reviewed, however, is enough for the charge not to be applicable that they are only splitting jealous hairs.

Each individual contributor has had his integrity respected. Where differences of convention appear in the spelling of non-European names they have been left alone. And where differences of opinion have appeared in non-factual matters no attempt has been made to reconcile them.

The case against von Däniken has several aspects. Some of his 'facts' are false and his arguments from these 'facts' are not acceptable arguments at all. In the matter of 'facts', for example, Mr. Gordon Whittaker shows that von Däniken appears to show great ignorance of the mass of real information available about South American antiquities.

Where von Däniken says that the Olmec heads are immovable, Whittaker is able to indicate that one was moved thousands of miles quite recently. Where von Däniken says that the Easter Island statues could not have been carved by the unaided inhabitants of the island, Mr. Whittaker is able to demonstrate that they were indeed so carved.

The 'mysterious' map of the Turkish Admiral Piri (or Piri Re'is) is shown by Dr. A. D. Crown, most circumstantially, to be a very impressive map, but not in the least mysterious.

Three writers, Professor Hennessy, Dr. Baker, and Mr. Houlsby, all find fault with von Däniken's Egyptology. Where von Däniken asserts that some pyramid stones weigh 12 tons each and were therefore humanly unworkable, the answer comes that all Egyptologists know that they weigh only 2½ tons. And Mr. Houlsby (who ought to know) says that there were and are no great engineering problems associated with pyramid building.

These are only examples of details which build up into a strong implication of carelessness by the author of *Chariots of the Gods?* It is not a 'stereotyped pattern of thought' that von Däniken has to fear, as he claims, from 'the classical method of research into antiquity' (p.70) but simple, hard, intractable, and publicly-available *facts*.

With regard to his method of argument, it is difficult for the editors, and for the contributors, to find any reason to respect it. They are unable to discover that he uses any of the commonly acknowledged techniques of intellectual advancement.

Conventional argument, of course, may be one of the 'lines upon which we have thought until now' and which we must leave if we are 'to set out on the arduous search for truth' (p.74). If so, and if existing methods of proceeding, like the syllogism, or making and testing hypotheses, or the mathematical assessment of probabilities, are all insufficiently arduous for the best future of archaeology, then it would be reasonable to expect from von Däniken a demonstration of how to proceed even more strenuously. That he has not provided it, and what he has in fact provided, are both shown by most contributors in passing, and explicitly by Mr. Nicolson and Mr. Grouse.

However, it is likely that in the field of human psychology Erich von Däniken is an intuitive genius.

The success of his book is not due to intrinsic merit. His case would have fallen on cold ears in 1772, and it is having a cold reception amongst the Himalayan sherpas at the moment. For so many unfactual facts and for so much elusive argument to be convincing there has to be a readiness in the readership, a prior willingness to be convinced.

Exactly why, at this time and in certain countries, the readiness is there is a question of great complexity. Two of the present writers are directly concerned with it: they see von Däniken as a clever man who has been able to tap the deepest anxieties of technological man. Several other contributors touch on the same possibility, and glance at radical deficiencies in public education which leave even 'educated' people vulnerable to extraordinary mythologies.

What has so far been said in this introduction could well be summed up in a quotation from Bergan Evan's *The Natural History of Nonsense:*[1]

> "Fallacy is always the product of certain processes in popular thinking: of arguing from negatives and analogies, of making false generalizations, or worshipping coincidence, of taking rhetoric for fact, of never questioning, or even perceiving the underlying conceptions that make for prejudice, and, above all, of a romantic delight in the wonderful for its own sake. And once made, the error . . . is likely to owe its vitality to intellectual currents and social forces with which, superficially regarded, it has no seeming connection".

Positively, there is one more thing to be said.

The Introduction to *Chariots of the Gods?* begins with words which are becoming famous: "It took courage to write this book, and it will take courage to read it." Is it not better to say that it takes *more* courage, and is more in accordance with the best in man, to learn to live patiently with mysteries than to escape them by inventing or accepting a sudden, simplistic and simultaneous solution?

We think so, and decline to trust in Chariots.

1. Vintage Books, New York, 1958, p. 251.

Archaeology of the Ancient Near East

PROFESSOR BASIL HENNESSY

J. B. Hennessy, B.A., D.Phil., F.S.A., is the Edwin Cuthbert Hall Visiting Professor of Near Eastern Archaeology at the University of Sydney. He was formerly Director of the British School of Archaeology, Jerusalem.

The recent publication and filming of Erich von Däniken's *Chariots of the Gods?*[1] has certainly created quite a stir amongst the general public and as it seems, somewhat indirectly, to involve the world of archaeology, a brief appraisal of some of the author's archaeological references and claims may not be out of place.

Many years before von Däniken's bleats for more accurate methods in archaeological research the world of archaeology had commenced its own clean up and today, can count on the unstinted assistance of all the empirical sciences. However, before archaeology can take full assistance from these new aids the basic collection and documentation of the material is in dire need of an improved methodology — certainly von Däniken's fascinating collection of unsupported, undigested, disconnected and often inaccurate claims does nothing towards the realisation of this desired end. There is no point in calling for technical assistance if one has already chosen to ignore the findings.

Let us take a look at some of von Däniken's 'facts' — a piece of cloth from Helwan (p. 43) which would apparently require a special modern fabric factory and experience — rubbish! — had the author wished to be sensational he might have gone back another 2,000 years to the well attested weaving of early Anatolia.[2] Spinning and weaving are amongst the earliest of the arts the Near East has bequeathed to us and from the neolithic period onwards we have ample evidence not only of the cloth, but also the looms and methods of manufacture — methods which have been studied widely by a legion of experts in the field.[3]

The curious collection of facts on page 42 of *Chariots of the Gods?* seem to be intended to demonstrate the assertion that ancient Iraq had a long history. The point could have been made more easily and more accurately with a reference to any of the dozens of excavation reports, general syntheses or histories of the area.[4] What von Däniken does not tell us is that between his earliest cited evidence and the '*sudden* appearance' (p. 43) of the Sumerians there is a well documented series of settled communities and cultures from 9000 B.C. onwards, culminating in the urban civilisations of the Mesopotamian region around 3000 B.C. The earliest definite record of anything Sumerian is at this time, though it seems likely in the demonstrable community of technology and other cultural elements that the 'Sumerians' may have been in the area for at least 1,000 years before this period.[5] The growth to full urban civilisation was a slow and laborious process and there is certainly nothing remotely startling in the growth of Sumerian civilisation — indeed as one looks back into those areas of our past history with modern eyes, one constantly wonders what took them so long!

Perhaps the most ill used of all man's early compositions, the Pyramids of Giza, are again called on as an inexplicable sudden appearance (pp. 96 ff.). It seems always to be the great pyramid of Cheops (Khufu) which attracts attention — To see no pyramids but those of the Giza group, with nearly all the attention concentrated on the Great Pyramid alone, means almost inevitably to produce in the mind the idea that this group of three pyramids is something entirely exceptional in Egyptian thought and practice, and that the Great Pyramid, in particular, is so absolutely unique as to render the commonplace theory of its having been merely a tomb very difficult of acceptance. What ought to be realised therefore, first of all, with regard to the pyramids is that the Giza group, far from being exceptional, is only one group out of many which still exist in Egypt, and many more which once existed but have now perished; that groups of pyramids once stretched, and still stretch, though with gaps now in the line, from Abu Rawash, five miles north of Giza to Meroe deep in the Sudan, between the Fifth Cataract and the Sixth, and within a short distance of Khartum; that every single one of these pyramid groups is part of a necropolis, of which it is usually the centre; and that everyone of these pyramids had in its time a temple attached to it and where offerings were made and regular ritual gone through with on behalf of the dead person whom the pyramid commemorated and whose resting place it was. It will then be understood that the pyramids, wherever they may be found, and whatever their size, were nothing more or less than tombs and that all the efforts of the ingenious individuals who have striven, with labour almost as vast as that which built the pyramids themselves, to show that the Great Pyramid is a divinely inspired compendium of everything that man has known, may know, or ought to know, as repository of all sorts of standard measures, an observatory (surely the most cumbrous and clumsy ever constructed), and a prophecy of all that has happened since it was built and all that will happen till it tumbles down, are simply 'vanity and a striving after wind.'[6]

Here it will be necessary to be laborious to refute von Däniken's Egyptian inventions. The Great Pyramid is composed of an estimated[7] 2,300,000 blocks of stone weighing on an average 2½ tons. The astronomical mathematics which arrive at 31,200,000 tons for the total construction (p. 102) we are only left to guess — perhaps the calculation was done under a force of 5 gs.

The structure is of limestone (soft) not granite (hard) as we might be led to believe (p. 85). The choice of the site which appears to worry von Däniken so much (p. 79) is perhaps explained by the following — 'As an example of stone having been quarried on the spot where it was required, that for the Giza pyramids may be mentioned. The stone of which the greater part of these pyramids is built is very characteristic, being highly fossiliferous and containing innumerable nummilities, and is identical with that of the plateau on which the pyramids stand and several of the large depressions nearby are the quarries from which this stone was obtained — The hollow in which the Sphinx is, for example, being one of the quarries.'[8]

Von Däniken's 'non-existent' rope, (p. 101) can be seen in quantities in the Cairo and other Museums, is illustrated and reported from a great many excavations and there are rope making scenes depicted on tomb walls of the Fifth Dynasty.[9] Light and heavy ropes were even in use many hundreds of years before the construction of the

'Great Pyramid' and have recently been found in First Dynasty tombs at Sakkara.[10] The (non-existent) rollers are equally well attested[11] as is the (non-existent) grain,[12] while the (non-existent) huts were found by Sir Flinders Petrie to the west of Kephren's pyramid.[13]

Yes, the ancient Egyptians did import wood (p. 97) and from at least nine different countries and in large quantities.[14]

As to the question of means and implements it may suffice to say that neither at the Great Pyramid nor anywhere else in Egypt has the least shred of evidence been found that the ancient Egyptians ever, at any period of their history used any mechanical appliances except the lever, roller and the inclined plane. The various dodges which have been attributed to them and confidently sketched in plan and elevation never existed, save in the brains of those who have drawn them . . ., on the other hand, Petrie found ample evidence of the use of long copper saws, at least nine feet in length which were probably used to cut the great blocks of stone . . .'[15] But enough of this — the considered opinions of those technologists after whom von Däniken howls can be read in a library of volumes which deal with the technical aspects of the construction.[16]

One can only hope that it is a coincidence that 1,000 million times the height of Khufu (Cheops) pyramid is approximately the distance between the earth and the sun (p. 99). If this is intended as an example of the visiting astronauts astronomical mathematics then it must have been sheer chance that brought them to earth. The original height of the pyramid was 481.4 ft.[17] not 490 ft. (p. 102).

Von Däniken may well ask why the Egyptian priests lied so shamelessly to the traveller Herodotus (p. 102); but we may equally well ask why the modern Herodotus with all the contrary facts before him should so shamelessly continue the deception.

Sumerian Gods *were* considered and represented in anthropormorphic form[18] (p. 42); granite was cut and used, but sparingly in the early Egyptian Dynasties[19]

(p. 5); the area of the base of the pyramid of Khufu divided by twice its height does *not* give π (p. 99). The stone blocks of Khufu's pyramid were *not* joined to a thousandth of an inch (p. 100); but why go on!

Perhaps the most relevant item of von Däniken's evidence is on page 42. There has not, to the author's knowledge, been any shred of evidence to suggest that coprolites (fossil dung) found in a pit at the proto-neolithic site of Tepe Asiab[20] were anything other than human; but if another origin is sought, to be consistent with von Däniken's other archaeological claims it could only be Bovine.

REFERENCES

1. VON DANIKEN, ERICH, *Chariots of the Gods?* London, Corgi, 1971.
2. MELLAART, JAMES. *Earliest Civilizations of the Near East.* London, Thames & Hudson, 1965. fig. 57 and p. 105.
3. LUCAS, A. *Ancient Egyptian Materials and Industries,* London, 1959. pp. 165-172.
4. MELLAART, JAMES. *Op. Cit.*
5. MALLOWAN, M. E. L. *Early Mesopotamia and Iran.* London, Thames and Hudson, 1965. p. 72 ff.
6. BAIKIE, JAMES. *Egyptian Antiquities in the Nile Valley,* London, 1932. pp. 104-105.
7. EDWARDS, I. E. S. *The Pyramids of Egypt.* Edinburgh, Pelican A168, 1961. p. 116.
8. LUCAS, A. *Op. Cit.* p. 68.
9. LUCAS, A. *Op. Cit.* pp. 160-161. DAVIES, N. de G. *The Mastaba of Ptahhetep and Akhethetep,* I, pl.XXV.
10. EMERY, W. B. *Archaic Egypt.* Edinburgh, Pelican A462, 1961. p. 223.
11. EDWARDS, I. E. S. *Op. Cit.* p. 266; and for general problems of construction and transport, *ibid.* pp. 254 ff.
12. EMERY, W. B. *Op Cit.* p. 237.
13. EDWARDS, I. E. S. *Op. Cit.* p. 155.
14. LUCAS, A. *Op. Cit.* pp. 488-500.
15. BAIKIE, JAMES, *Op. Cit.,* pp. 116-117.
16. EDWARDS, I. E. S. *Op. Cit.* pp. 304-306.
17. *Ibid,* p. 118.
18. MOSCATI, SABATION. *The Face of the Ancient Orient,* London, 1960. pp. 27 ff.
19. LUCAS, A. *Op. Cit.* pp. 72-74.
20. MELLAART, JAMES. *Op. Cit.* p. 22.

Einstein and the Spaceman

COLIN GAULD

C. F. Gauld, B.Sc., Ph.D., Dip.Ed., is a lecturer in Science Education at the University of N.S.W. Besides being an avid reader of science fiction and adult fantasy, he has shown for years in his conversation, lectures and writings a deep interest in the history and philosophy of science, in astronomy and cosmology, in the relation between scientific knowledge and Christian faith. He has had wide experience in teaching, both in Australia and the United Kingdom. His original research was done in the field of Nuclear Physics.

Chariots of the Gods? is a book in praise of visionaries (p. 90), of daring speculation (p. 16)[1] and of fantastic theories (pp. 46, 151, 154). It supports these not because they are true (pp. 72-3) but because 'all things are possible' (pp. 47, 119). It condemns as narrow-minded and reactionary (pp. 45-6) those who, on other grounds, reject the visions which take so much 'courage' to present.

Von Däniken thus sets the scene for his own speculations accompanied by appeals for serious consideration (pp. 74, 95, 180) and for expensive research commitments (pp. 112, 118, 119, 132, 179, 180). Is his characterisation of the place of speculation in the growth of knowledge a correct one? Are we always better off if we accept the most widely improbable theories because they *could* be true? Are these speculations always proved correct in the end? If an ancient Greek scientist had come up with the theory of relativity would his contemporaries be correct to accept it or reject it?

Scientific knowledge only grows and bears new fruit when there exists the possibility of confirming or denying predictions which follow from some theory. The ideas of Herakleides,[2] Aristarchus[3] and Bruno,[4] imaginative and daring though they were, remained sterile because there was no way of testing them. Their contemporaries had every right to reject them and were able to clearly state their reasons — there was little evidence to support the ideas and plenty on which to base the alternative view.

It is thus quite untrue to say, as von Däniken does, that scientists are concerned with things only when they are proved (p. 45). What is required are predictions which can be tested. Calculations and experiments are often carried out to see if something is true or false. For example, Einstein's theory of relativity predicted that if light, travelling from a star to earth, passed near the sun it should be bent by the sun's gravity. During a total eclipse of the sun in 1919 an expedition from England travelled to the South Pacific to see if this prediction was true or not (Jones, p. 196).[5] Many scientists today are searching for subatomic particles which, if some current theories are true, should exist. If the theories are not correct the search will probably be unsuccessful.

The reaction of research workers to theories of reputable scientists like Sagon and Shlovsky about Phobos is certainly not to accept speculations that Phobos must be an artificial satellite and that Mars once possessed an advanced civilisation (p. 154). At this stage these views are completely untestable and so must be ignored as were the theories of Herakleides, Aristarchus and Bruno. The first of these turned out to be unacceptable, the second is similar, but not identical with the one we believe to be true today and the third is still not open to experimental investigation. Research workers accept the data regarding the motion of Phobos and continue in their attempt to account for it with theories that are able to be checked. They would be foolish to do otherwise.

Is Nothing Incredible?

In his book von Däniken makes some remarkable statements about speculation which he openly contradicts in other sections. For him everything is credible (p. 47), and not only credible but possible (p. 119). One gets the impression that, from the examples of tachyons, neutrinos and antimatter (p. 141), all the scientist has to do is imagine something not contradicted by present day theory and he will inevitably find it. The existence of tachyons (that is, particles travelling faster than the speed of light) has been suggested by a number of scientists (Feinberg) and denied by others. It is not a question of just showing they could exist but of finding out whether they exist or not. For many years it was believed that irregular movements of the planet Mercury were caused by the gravitational effect of another planet much nearer the sun than Mercury (Smart, p. 49). This 'unknown' planet was labelled Vulcan but, in spite of intensive searches, was never found. Einstein's theory of relativity finally explained the irregularities in the movements of Mercury and the existence of Vulcan became 'incredible'. Even von Däniken does not really believe that everything is possible but accepts the necessity of deciding on the basis of evidence what to believe and what to disbelieve. For example, in discussing the origin of the crater in Siberia (p. 150) he rejects the laser theory because it is too incredible and the anti-matter theory for reasons which are outlined. The theory that the constructions in the Palpa Valley (p. 31) are Inca roads is ruled out as preposterous. These examples, among many others, show von Däniken's inconsistency. He does *not* believe that everything is credible or possible even if Thomas Mann did (p. 119).

While he applauds speculation beyond the bounds where it is testable (pp. 72, 73, 74) he does offer evidence for his own views, so we must turn to this in order to assess the quality of his views.

Much of the material is of an archaeological or theological nature and will be discussed in other sections of this book. However von Däniken takes great pains to add scientific weight to his thesis in spite of his stated mistrust of academic scholars (pp. 28, 33, 34, 45, 46, 90, 107).

Von Däniken's Nonsense Numbers

Numbers are impressive even if we don't understand what they mean, but the statement on page 11 about fifteen digit numbers, considered important enough to include in the introduction to the book, is utter nonsense. Of course these can be registered by computers. The list of figures on page 125 again looks impressive but how it is supposed to give any idea of the accuracy of the Mayan Calendar is difficult to see. From this list two conclusions can be drawn. The first is that the Mayans could multiply 20 by 18 by 20 by 20 by 20 by 20 by 20 by 20 and come up with the correct answer of 23,040,000,000. (What the small subscripts 2 and 1 in the answer are supposed to mean is not explained). The second is that if this is a calendar then one 'tun' appears to be the same as a year and 360 days is not very accurate at all. These figures seem at odds with the figures on page 75 which are also related to the Mayan Calendar. How are we to conclude anything about its accuracy from the data given?

Green Bank Formula

The Green Bank Formula (pp. 165-7) discussed at length in *Chariots of the Gods?*, contains a number of factors which, if they were known and multiplied together, would give the number of civilisations in the galaxy able to communicate with one another and with us. Von Däniken states on page 165 that this formula gives us a result of 50 million civilisations — an impressive figure. On page 166 he states that this is only the maximum number and that the minimum possible value is 40 — not quite so impressive but still remarkable. Walter Sullivan in his discussion of this same formula (Sullivan, Chapter 17) discloses the uncertainty surrounding the value of the factor labelled L. This is the time span during which a civilisation has the ability to communicate across space and not, as von Däniken states, the life span of a civilisation. For human civilisation, the period referred to by the factor L began no earlier than about 1960. The question to be first answered is 'when civilisations become capable of destroying themselves as we are now how long are they likely to last?'. The only real answer is 'who knows?'. The values chosen at the conference for L and based very much on guess work give for the minimum number not, as von Däniken claims, *more* than 40 civilisations but *less* than 20 (Sullivan, p. 285). It is also significant that at a conference dominated as it was by physicists and astronomers (Sullivan, pp. 268-9) the two biological factors dealing with the fraction of suitable planets on which life had actually emerged and of these the fraction on which intelligent life has developed were considered to be 100% (Sullivan, pp. 274, 277). The views of some evolutionary biologists that at least the second factor could be very small indeed was not considered by members of the conference (Sullivan, p. 275). Using these figures, the final value would be more like one civilisation, namely those living on earth. The problem with the Green Bank Formula (or with similar calculations on pp. 16-17) is that no one really knows the value of most of the factors which have to be multiplied together to get the answer (Sullivan, p. 287).

Did Gulliver Travel?

Von Däniken expresses surprise that not only did Jonathan Swift mention in *Gulliver's Travels* that Mars had two moons 150 years before they were discovered but that he was able to give precise data about them (pp. 152-3). What he does not tell us is that Swift's figures were incorrect. Instead of taking 10 and 21½ hours to orbit Mars, Phobos and Deimos take 7½ and 30 hours (Sullivan, p. 184). Instead of orbiting at distances of 12,600 and 21,000 miles from Mars, Phobos and Deimos move at distances of 5,800 and 15,000 miles respectively (Sagan and Leonard, p. 193). These are the figures that von Däniken calls 'precise'.

Von Däniken's Inconsistency

A number of other examples show a lack of understanding of the scientific data appealed to. For example, he is happy to quote Einstein's theory of relativity to support his case in some places (pp. 22-3, 165) without realising that in doing so he undermines his argument in other places. Belief in the theory of relativity rules out the possibility of a form of energy which can be transmitted everywhere simultaneously (p. 164). If von Däniken's form of energy really exists then we cannot trust those sections of his book where he refers to the theory of relativity. On page 155 reference is made to *Worlds in Collision* by Emanuel Velikovsky and it is stated that Velikovsky believed the planet Venus was formed when a large comet crashed into Mars. It is claimed that confirmation of Velikovsky's theory by data from Mariner II lends support to von Däniken's thesis (p. 155). However, Velikovsky clearly stated that he believes Venus originally came from Jupiter and, after undergoing a number of collisions with Mars and the Earth (Velikovsky, p. 361), settled down into its present orbit during the latter part of the eighth century before Christ (Velikovsky, p. 199). Von Däniken's lack of understanding of Velikovsky's theory not only shows through his erroneous description but also because he does not realise that if he appeals to this theory for support he must reject some of his earlier data.

Velikovsky claims that in the encounters between Venus, Mars and the Earth, the lengths of the day, the month and of the year changed a number of times in the 2,000 years before the birth of Christ (Velikovsky pp. 125-9, 300, 342).

If Velikovsky is correct how then can we know what the heavens were like 27,000 years ago (p. 35), or where the constellations were 10,000 years ago (p. 43, or whether the Egyptian Calendar from 4,221 B.C. (p. 86) was accurate or not? If Velikovsky is correct, the Great Idol of Tiahuanaco (pp. 34-5), the cave drawings of Kolistan (p. 43) and the Egyptian Calendar (p. 86 are irrelevant pieces of data.

Part of Velikovsky's evidence for this theory rests on the identification of the planet Venus with the ancient Mexican god Quetzalcoate (Velikovsky 158-9) and this, too, contradicts von Däniken's speculations about Quetzalcoate in which he is apparently seen as a visitor from space (pp. 127-8).

When he shows so much misunderstanding of such easily accessible evidence as that described above it becomes senseless to accept von Däniken's interpretation of data which is more difficult to follow up.

Subjective Selection

In *Chariots of the Gods?* von Däniken has a lot to say about academic scientists and his attitude is quite clear cut. Those who offer information which appears to support the thesis of the book are treated with respect (p. 141). The caution of Shapley and Ley (p. 16) is noted, the daring assumption of Miller (p. 16) is used in evidence. Bellamy's and Allan's close reasoning is mentioned (p. 34), the fantastic theories of Sagan and Shlovsky (p. 154) are described and those who represent the new study of exobiology (pp. 167-8) are credited with wishing to break through taboos and smash walls of apathy.

The ideas of the mostly anonymous people he disagrees with are dismissed, often with sarcasm supported by little argument or because von Däniken just can't believe things happened the way they said it did.[6] Others are just ignored. Soon after the Green Bank Conference, dominated as it was by physicists, G. G. Simpson, the Harvard paleontologist, published a book (Sullivan, p. 275) in which he deplored the amount of money spent on exobiology (the study of life in outer space) and supported his position by outlining the arguments from evolution for the *improbability* of finding intelligent life beyond the Earth. Simpson's views or those of biologists H. F. Blum and M. Ycas are not mentioned by von Däniken. His reference to the theories of Einstein and Velikovsky have been mentioned before and show the same selectivity. He uses those sections which appear to strengthen his thesis but remains silent about details which would undermine his case. This inconsistency borders on academic dishonesty!

Spacemen of the 'Gaps'

We on Earth are in the early stages of a space travel program which may or may not develop very far beyond our solar system. Seen in this long range perspective our space technology is primitive and if we were to achieve those goals envisaged by von Däniken (pp. 22-3) one would expect radical changes to have taken place in our technology including presumably the life support and communication equipment of our space travellers. Yet the God in the chariot from space has been created very much in the image of our own primitive space traveller and technologist of the 1970s,[7] the photographs and drawing Tassili (opposite p. 97), the drawing of the stone relief inside the Temple at Palenque and the comparison photograph of American astronauts ready to take off.

Von Däniken's spaceman can best be described as a 'spaceman of the gaps' since anything at all which at present is an unexplained gap in our knowledge can be used as evidence for 'his' existence.[8] Almost any mystery, and there is a large number of fascinating ones in any area of scientific knowledge, can be easily explained by appeal to the superior spaceman hypothesis. However, if we lived in an earlier age many of the mysteries which we may have then attributed to our hypothetical spaceman would now quite satisfactorily be accommodated by normal scientific theories and the gaps in our knowledge would gradually close, squeezing out this 'visitor from space'. This will probably be the same with the data von Däniken, and probably many professional archaeologists and scientists, find so difficult to understand now.

BIBLIOGRAPHY

BLUM, H. F. On the Origin and Evolution of Living Machines. *American Scientist,* Volume 49, December, 1961. pp. 474-501.

BRUNO, G. On the Infinite Universe and World, in *Theories of the Universe,* by M. Munitz. Free Press, 1957. pp. 174-183.

FEINBERG, G. Particles that go Faster than Light. *Scientific American,* February, 1970. p. 68.

JONES, H. S. *General Astronomy*. Lond., Edward Arnold, 1951.

SAGAN, C. and LEONARD, J. N. *Planets.* Amsterdam, Time-Life International, 1967.

SIMPSON, G. G. *This View of Life.* N.Y., Harcourt, 1964. [Especially chapter 13, "The Nonprevalence of Humanoids", which is reprinted in the journal *Science,* Volume 143, Number 3608, 21 February, 1964. p. 769-775.]

SMART, W. M. *The Riddle of the Universe.* Lond., Longmans, 1968.

SULLIVAN, W. *We Are Not Alone.* Harmondsworth, Penguin, 1970.

TOVLININ, S. and GOODFIELD, J. *The Fabric of the Heavens.* Harmondsworth, Penguin, 1963.

VELIKOVSKY, E. *Worlds in Collision.* Lond., Gollancz, 1950.

YEAS, M. *Are We Alone in the Universe?* in "Light and Life in the Universe". S. T. Butler and H. Messel (eds.). Syd. Shakespeare Head Press, 1964. pp. 217-241.

REFERENCES

1. The page numbers in brackets refer to the Corgi paperback edition, 1971 of *Chariots of the Gods?*
2. Herakleides, a contemporary of Aristotle, introduced a theory of the universe quite opposed to the most commonly accepted at the time. He suggested that the earth travelled around the sun and the other planets moved around the earth. Interesting but wrong! See Toulmin and Goodfield, *The Fabric of the Heavens.* pp. 131, 203-205.
3. Aristachus, in about 280 B.C. suggested that the sun was at the centre of the universe and that the earth and all known planets travelled around it. This view was virtually ignored for 1700 years until Copernicus re-introduced it. (Toulmin and Goodfield. pp. 132-138.)
4. Giordono Bruno in 16th cent. speculated that there were many worlds populated by intelligent beings.

5. See select Bibliography for reference books quoted (at the end of this section).

6. *Chariots of the Gods?* pp. 28, 31, 33-4, 37, 44-6, 49, 89, 90, 96, 97, 101, 107.
7. *Op. Cit.* pp. 24, 32, 42, 48, 78, 83, 123, 124, 131, 150, 151.
8. *Op. Cit.*, pp. 27, 42, 43-4, 47-50, 88, 97-102, 105, 113-7, 120-131.

The South Sea Bubble Burst

A. D. CROWN

A. D. Crown, M.A., Ph.D., is Senior Lecturer, Department of Semitic Studies, University of Sydney. He is also the Federal President of Australian Friends of the Hebrew University and a founding member of the Academy of Forensic Sciences.

> 'What has happened will happen again and what has been done will be done again and there is nothing new under the sun. Is there anything of which one can say, "Look, this is new"?'

There can be little doubt that a group of responsible scholars drawing on their training can demolish the arguments put forward by von Däniken in his *Chariots of the Gods?*, for there is barely a statement in the book which will stand scrutiny. More important than this examination, is some appreciation of why this book has had a stunning impact on so many readers and why, in a world which spends so much of its wealth on education, so much trust has been placed on patently false statements. What is the common factor in the life and expectation of so many people that has made them willing to accept the unusual attitudes in this book?

The History of History

Paradoxical as it may seem, the answer is indicated by the attitudes current today to the view of Koheleth, the preacher, quoted above. We live in a world which has digested at a gulp the marvels of interplanetary travel as an expected bi-product of the application of human inventiveness to mechanisation. Change, whether it be in the means of communication via radio, television, or earth satellites, or whether it be in social attitudes spread over our increasingly monocultural planet, is the hallmark of our day. Everything seems new, nothing appears to be static or to have roots. By way of contrast, ages past seem to mark little more than an infinitesimally slow climb from leopard-skinned barbarism to our present level of technological expertise. This view of history is fostered and encouraged in the schools of the civilised world by the somewhat unrealistic view which divides the ages of mankind into a sequence based on the cultural development of Europe and Britain. In this scheme history began with the Roman expansion northwards and their conquest of the Ultima Thule. Gaul, Britain and the lands to the east were occupied by demi-men, woad-dyed and untutored. The chthonic darkness of European paganism was illuminated only by the spread of Christianity. Such is the conceit of European historians that they have dubbed the ages between Lindisfarne and Wittenberg, the Middle Ages, ignoring entirely the fact that civilization was old when the Romans were a young people and that the culture they diffused to the north lands was a latter day development, a Johnny-come-lately culture in the history of human development. The Romans were the catalysts, the social pivot, the human fulcrum around which the experience, inventions and order of the Near East were swung into their northern European ambit. In terms of recorded history, the Roman conquest of Britain marks the onset of the late Middle Ages, and the Near East had as long a prehistory as it had a history. It is the failure to recognise this simple fact by so many of our contemporaries that has made them prone to accept the shibboleths propounded by von Däniken. It is this single failure that serves as the common preconditioning factor in the success of *Chariots of the Gods?*. Without the realisation of the richness of life in the Ancient Near East, there is a natural failure to recognise that many of the processes which are essential to the manufacture of our mechanical culture symbols, radios, television sets, are dependent upon discoveries in mathematics, astronomy, chemistry, metallurgy and writing, that were made by our Sumerian and Semitic forebears.

Von Däniken devalues human ingenuity and inventiveness and denigrates the achievements of the civilizations of the Ancient Near East. He works upon the ignorant with the tools so carelessly given to him by European historians.

There is another aspect of the concern for novelty in our contemporary culture which illuminates further von Däniken's success in convincing people that our ancestors could not have achieved what they did without some miraculous intercession from interplanetary travellers. This is the failure to recognise that machines do not necessarily present mankind with his first opportunities to perform operations or even that machines are not necessarily power driven. Some machines make novel contributions to cultural progress; some few more, such as the gun, the internal combustion engine, the power loom and the bulldozer add a dimension of intensity or scale to functions that were provided by more primitive devices such as the bow, the horse, the hand loom and the spade. Most machines simply provide a low cost substitute for that most expensive commodity, labour. It is quite improper to suggest that because an operation is a large scale undertaking which, today, would need the application of heavy and expensive machinery, it was beyond the means of our forebears. We must remember that from the moment man began to live in cities (somewhere between 6,000 and 8,000 years ago) he was forced to rely on the co-operation of his fellows. Urban living needed co-operation to maintain food supplies to the concentration of people, to build and man walls against enemies, to provide water for irrigation and drinking, to build temples, to provide ships and wharves for the movement of trade goods and capital for the purchase of merchandise. This same co-operation which made cities viable made it possible to undertake large scale building projects, whether that project was the erection of a colossal statue or the damming of a river for irrigation purposes. The mystery of the movement of large stones, colossi and the like found throughout von Däniken's writing depends for its appeal on the latter day mythology about the nature of machines and our misunderstanding of the evolutionary nature of urban living.

Solution to the Map Mystery

There are other mysteries in the book which are also of the author's manufacture. Perhaps the most striking of these is that which relates to the map which is twice presented in pictorial form. This form alone appeals to the reader for it is a common conceit of mankind that his eye cannot be deceived; seeing, it is said, is believing. The map is said to have belonged to an officer of the Turkish Navy, Admiral Piri, or Piri Re'is (= admiral), and is said to have been discovered in the Topkapi Palace in the eighteenth century. Its age is not clearly stated but is implied, a technique most commonly used by the author apparently as a means of avoiding making patently false statements. The implication of the date and authorship of the map appears in the words on page 31, 'How are we to explain that? Should we be satisfied with the legend that a god gave them to a high priest? . . . Or should we boldly stir up a wasps' nest and claim that this cartography of our globe was carried out from a high-flying aircraft or from a space ship?' Nowhere is there any statement as to who claimed that the Piri Re'is map was given by a god to a high priest and there is certainly no such statement on the map itself. One must assume that von Däniken is here setting up his own straw man for summary execution. It is not certain what von Däniken means by the plural 'them' in the quoted sentence. On page 29 he speaks of two atlases in the Berlin State Library with maps by Piri Re'is of the Mediterranean region, and it may be that these maps are the ones included in his 'them'. It may also be argued that von Däniken includes the projection of the Piri Re'is America's map on to a modern grid in this plural. Support for this view is given by his statement that the Piri Re'is map was made by aerial photography from a centre over Cairo, a view which could only hold if he relied on the azimuthal projection he pictures as the source of his information. An essential part of von Däniken's argument is that the maps are accurate and that they show parts of Antarctica known today only by echo-sounding techniques.

The Piri Re'is map shown by von Däniken can also be found as plate LVII in Leo Bagrow, *History of Cartography*[1]. According to Bagrow, this map was not discovered in the eighteenth century as claimed by von Däniken but was found in the Topkapi Seraglio in 1935; a detail, but not atypical of the many inaccuracies in the *Chariots of the Gods?* The surviving map is only a fragment of the original and shows West Africa, parts of the Indies and South America. To check the claim that Antarctica appears on the map one must examine the fragment carefully. This first datum of importance is that there is a distinction made on the map between the northernmost part of South America and the southerly part shown on the map. Small details are noted of the northern part, rivers are marked and are identifiable and there are some indications of the nature of the terrain behind the coast. The coast is not stylised in any way. Every fragment is different and it is clear that some accurate tradition underlies this map. The coast has clearly been mapped or seen. The southern part of South America is marked rather differently. Here are one or two islands, but no rivers. There are one or two promontories, but for the most part the coast is stylised with the same 'battlement' being repeated. Here the coast has not been surveyed, and the 'battlement' is a filler to provide some sort of linkage for the two or three promontories that appear to have been sighted from well out to sea. There are no details of what lay inland behind the stylised coast, reinforcing the view that this coast was not surveyed. How far south was the surveyed area? On the assumption that the unsurveyed area truly shows a number of promontories sighted from the sea and that the directions shown in the unsurveyed area are generally correct on the basis of these sightings, one must find some place on the South American coastline where there is a marked easterly projection and which lies to the south of a coastal plain buttressed by a mountain range. Those conditions are indeed met between the equator and 5° South by the coast of Brazil as far as Cape de Sao Roque. Careful examination of the explored coast bears out this conclusion. The archipelago at the southern end of the explored portion of the coast are the islands of the mouth of the Amazon. The great cape which forms the southern most point on the map is Cape Sao Roque itself. Thus, despite the claim of von Däniken, Antarctica, echo soundings or no, is *not* shown on the map.

In any event, the authorship of the map is well known. Admiral Piri claimed that he had drawn it himself basing his work on older charts including some made by a sailor who had sailed on three voyages with the great Christopher Columbus. No wonder the map is tolerably accurate; it is one of the copies of the cartography of the master explorer. Not only have we no need to postulate a high priest for this map we have no need to postulate any especial prescience for the data assembled thereon. The data do not present a picture of anything but careful cartography of explorers who knew the coast of South America as far south as the Amazon, who had sailed eastwards into the Atlantic parallel to the Brazilian coast, but as yet, had not ventured to the south. That venture was to come in 1519, six years after Piri drew his map, when Magellan found the straits that today carry his name.

There should be even less surprise that the atlases in the Berlin State Library carry accurate maps of the Mediterranean and the area also dating from the time of Piri. The sixteenth century Turkish cartographers were amongst the best of their contemporaries. Turkey was building up its sea power and was seeking to dominate those seas which were in reach of its fleet. Piri records that he culled information from ships plying the Red Sea and the Indian Ocean as well as the Mediterranean. In any event, von Däniken again undervalues the achievements of ancient mariners who, in their *periploi,* left admirable sailing guides and instructions as to the anchorages and harbours of the Near East, a millenium before Piri.

There is less to be said of the modern azimuthal projection. This is a projection of the whole world. The Piri Re'is map is a fragment. Thus whatever else it might be the azimuthal projection bears no relationship to the Piri Re'is map or maps. Even if the extant parts of the Piri map were put onto a modern azimuthal projection they could show nothing of the distortion which von Däniken makes so much of, as a feature of aerial photography. Distortion at the peripheries is one of the features of azimuthal maps whereas at the centre they tend to show true direction and give a reasonable impression of a small area. Since the Piri map extends only to 5° south of the equator there would be no distortion if it were to be projected on to an azimuthal grid based on or near Cairo. The projection in von Däniken's book is a modern one drawn in 1957 and culled, no doubt, from Charles H. Hapgood, *Maps of the Ancient Sea Kings,*[2] though Hapgood makes no special claims for the projection. All the Azimuthal grid shows is that whoever drew it had a sound training in modern cartography. It says nothing about space craft, Gods, or the Piri map.

REFERENCES

1. BAGROW, L. *History of Cartography.* Harvard University Press. 1964. Plate L. VII.
2. HAPGOOD, C. H. *Maps of the Ancient Sea Kings.* 1957.

Chariots in the Land Flowing with Milk and Honey

E. C. B. MACLAURIN

E. C. B. MacLaurin, M.A., B.D., is Head of the Department of Semitic Studies, University of Sydney. For many years has been specially interested in Canaanite and Ugaritic culture.

This book consists of a great mass of unrelated and unevaluated facts, semi-facts and non-facts. It is as discriminating as a lyre-bird's nest, and contains archaeological and scientific fragments collected without any guiding principle from sources widely scattered in space, time and reliability. It would not be a hard book to write once the author had arranged his card-index under appropriate headings, but it is a book almost impossible to controvert with any degree of expertise since such a task would demand a specialist knowledge in half a dozen different disciplines. This chapter will deal with two specific areas in von Däniken's claims, one in the Canaanite field and one in the Egyptian field; I will leave to the astronomers the implication (on p. 35) that the moon is only 27,000 years old and to archaeologists the statement (p. 110) that the pre-pottery Neolithic peoples of Jericho knew how to make plaster of Paris.

On page 41 von Däniken takes the phrase 'kingship came down from heaven', so common in ancient thought to express the idea that sovereignty is the delegation of power by God, as meaning that other beings—presumably not everliving—came to this earth in a spaceship and were hailed by the primitive inhabitants as divine. This, of course, solves no philosophical problems for it simply pushes the source of creation and human inventiveness one step further back; it shares the error commonly held today that because early man wore skins and used stones as tools he was also intellectually primitive. In truth the reverse may be the case and early man, freed from time-and-energy-consuming activities such as hedge us about, may have had much more opportunity to meditate deeply than we do. The achievements of the last century, made so far as we know without any direct intervention from outside human capacity, should warn us to hesitate before we say of any skill ascribed to early man 'Impossible'.

The New Breed

Von Däniken associates the idea of extra-terrestrial beings visiting the earth with the biblical statement (Gen 6:2; Job 2:1 etc.) that there were such beings as Sons of God and that they were capable of copulating with and impregnating human females. The proposition is scientifically untenable, for it would presuppose that the Sons of God had a carbon-based physiology and were physically adapted for entering the human female and that the species were sufficiently close to permit the chromosomes to be interchangeable — a circumstance that does not apply when bestiality takes place between beings both basically descended from the same primitive terrestial stock. Our kinship with other mammals is clearly demonstrable, but — except in the case of certain high primates and then only under special laboratory conditions —our ova and sperm cannot inter-fertilize with theirs. Yet von Däniken speaks of a special race born from this divine-human connection which is superior to all other races and which is predestined to surpass and survive.

The origin of this ancient legend about the Sons of God and the daughters of men is probably to be found in the growth of monotheistic religion from the early sea-pantheon of the fecund Eastern Mediterranean; here the ancient sea-god, the forerunner of Ugaritic Yam and classical Poseidon, dwelt with the Happy One, his wife, and their seventy sons. His sovereignity, his wife, and his family became the property of victorious El when the Semites overwhelmed the proto-Aegeans on the coast of Canaan, and then the El/Yahweh synthesis after Sinai transferred the carapace of the legend, stripped of its living flesh by the theologians of the new faith, to the God of the Hebrews.

The story cannot be taken literally as von Däniken has sought to do; equally fantastic is his view that human bodies equipped with horned animal heads are spacemen wearing wireless antennae. The practice of human beings dressed up as animals is a very ancient and very basic one; its antiquity may be seen in the Palaeolithic figures on the walls of prehistoric shrines in the caverns of France and Spain, its basic place in our psychology when you hear your little son running round on fours pretending to be a puppy. There is in us a deep feeling of kinship with the animal world which is the basis of totemism, and since this feeling of kinship exists it is quite natural for men to seek to gain the strength or guile of the animals he admires by sympathetic magic, by copying their appearance. And in addition such a disguise will conceal the hunter and mask his human smell as he approaches his quarry. The phenomenon can be so well explained in simple naturalistic terms that any attempt to interpret it in terms of beings from outer space is ludicrous.

Sodom and Gomorrah

The story of Lot and the destruction of Sodom and Gomorrah is also capable of a purely terrestial explanation. Von Däniken attributes it to an explosion of nuclear material set off by extra terrestial beings working to a time-table; it is much more likely to have been a traditional account of an explosion of an oil-field caused by earthquake. The Dead Sea on whose shores the unhappy cities stood is situated on the Great Rift, the longest fault in the earth's surface, which begins in the valley of Syrian Orontes, extends along the Jordan Valley to the Gulf of Aqaba, then across the Red Sea to the Sabie River in the Transvaal forking on the way eastward to the Gulf of Aden and westward to the central Congo. At some time movement in this fault has exploded an oil-field near the Dead Sea, and the extrusion of salt, bitumen and other phenomena associated with oilfields has taken place. The date of this natural event is impossible to determine but must have been later than Neolithic man's first attempts at building settlements, which would place it after the VIIIth Mill. BC. It is pre-Hebrew and may be connected with the natural horrors prophecied in the Papyrus of Ipuwer (XIIth Dynasty, beginning of IInd Mill. BC).

Similarly, (on p. 62) von Däniken claims that the course of the Flood is historically proved; this view is not shared by competent archaeologists or historians (see, for example, John Bright.[1]) I am inclined to agree with Sir Leonard Woolley[2] and von Däniken on this point more than most modern scholars are prepared to do. One of his remarks on biblical origins reveals von Däniken as a fundamentalist as well as a predestinarian Calvinist (as we have already seen). He enquires what language did Moses use when he wrote the Pentateuch; the belief that Moses wrote the

Pentateuch has long been abandoned on historical as well as literary grounds. Possibly quite a fair proportion of it may be traced to his genius, perhaps even to some written source, but it is quite impossible to retain the old belief in his authorship. We do know, however (and apparently von Däniken does not) that in the time of Moses, Canaanite was a literary written language, for quite a fair body of literature has been recovered from the ruins of Ugarit. The language of these texts is either Canaanite or Amorite or some dialect closely akin; it is sufficiently close to the Hebrew of the Old Testament for a good Hebraist to be able to work competently in it and it is written in an alphabetic script so close to the alphabet of Phoenicians, Hebrews, Greeks — and hence ours — that the question of the identity of the language of the early Pentateuch should never have been asked.

Von Däniken's statements about the civilization and religion of ancient Egypt also show serious gaps in his knowledge. He claims for example (p. 96) that ancient Egypt appeared with a ready-made civilization; this is quite contrary to the findings of Archaeological research. He would have been far closer to the truth if he had stated that the *origins* of Egyptian civilization are unknown and may have come from some hitherto undiscovered source but had undergone a long period of development in the Nile Valley. In any case, the basic structure was so basic that it would be ludicrous to attribute it to creatures culturally as advanced as spacemen must necessarily be supposed to be; it is probably far safer to follow the claims of Egyptian mythology and accept the probability that civilization, like agriculture, was brought from Syria and the Fertile Crescent (where the ancestors of most of our cereals originated) by some Osirian-like migrant. He refers to 'non-existent grain and ropes' (on p. 101) but in fact grain-growing has been practised in Egypt from early pre-Dynastic times, and the oldest paintings include some (e.g. Theban tomb 261), on which ropes are represented: in any case, the author has failed to indicate why the highly technological astronauts were still content with cord made of vegetable fibre, which is subject to decay and fire-damage when their space-ships were presumably made of some alloy so impervious to heat and friction that vessels made of it could travel at fantastic speeds for enormous distances and periods of time without it becoming so fatigued that a return-flight to its place of origin became impossible; if such 'ropes' did indeed exist then he would need to explain why the astronauts did not impart the secret of how to make them to their followers, especially if, as he wrongly asserts, these pupils had no ropes.

He asserts quite falsely that the Egyptians could not have imported sufficient wood to build the Pyramids (p. 97) by the methods usually attributed to them — the laying-down of vast quantities of parallel rails, made smooth with (olive?) oil along which the vast stones used in their Pyramids and other buildings were sledded. The first point to notice about this is that wheeled wagons were impractical for the great weights involved, and the Egyptians found their own method of transporting heavy material — whatever it was — so efficient that the wagon did not appear in common use in Egypt until after the Asiatics had introduced the chariot during XVIth Cent. B.C. We know, from contemporary paintings and reliefs in Egyptian tombs, that heavy goods were floated down the Nile on barges drawn by teams of slaves along the banks; these slaves used tow-ropes, and the method is so efficient that it was used up to the end of the XIXth Cent. A.D. We actually know the dimensions of some of these barges — thus Queen Hatshepsut's obelisk barge was 63 metres long and 21 broad; it was about 1500 tons displacement, built and used whilst the country was still in the Bronze Age. Innumerable funerary barges are depicted upon the walls of tombs and some of them actually survive in the tombs themselves. The sea-going ships of early times, both those of the Egyptians and of the Sea-peoples, are also depicted on the walls of tombs and temples, and Egyptian writings speak of ships of Byblos, the port where they were built and whence they plied their trade. Some of these pictures show barges, propelled by oars, carrying beams so long that they overhung both ends of the vessel. The land-bridge was heavily forested in early times, and so insatiable was the hunger of Egypt for wood (contrast p. 97!) that long before the settlement by the Hebrew invaders serious erosion had begun, to be hastened considerably by the folly of the Hebrews who actually cleared the forest from steep slopes and started the total devastation of the land as we see it today (Josh 17:15 ff). In pre-Hebrew times Canaan for over a millenium had been called in Egyptian literature 'a land flowing with milk and honey' (Josh 5:6) but Joshua's blood-thirsty vandals soon changed this (Josh 8:19, 24, 28; 9:15; 23; 10:10, 20, 26, 28, 30, 32; 11:11, 17, 20, etc.).

Light in the Tombs

Von Däniken points out that the painting of large areas of wall inside dark passages constitutes a problem for him. He states, quite correctly, that there are no signs of carbon from any sort of lamp or flare, and that electricity was unknown to the ancient Egyptians (quite irreconcilably, he states that the remains of an electric battery was found in Mesopotamia). He consequently concludes that the astronauts must have introduced to the ancient Egyptians some other form of artificial illumination; but if he were to travel to Luxor today I could introduce him to a septuaginarian dragoman Ahmed Yahya who has taken me into the inmost recesses of tombs in the Valley of the Kings, perfectly illumined without the aid of artificial lighting. Sheets of polished metal placed at given places along the passageways reflect the rays of the sun so perfectly, and the placing of these 'mirrors' is so skilfully arranged, that electric lighting is unnecessary.

Van Däniken's chief weakness is sheer ignorance; thus on p. 85 he states that the island of Elephantine in the Nile is so called because it resembles an elephant in shape, and that this can only be recognized from an aeroplane as there is no hill in the vicinity from which this can be seen. This ignores several facts: 1. the word *elephantinos* in Greek does not mean *elephant;* it means *ivory* and is a translation of Egyptian *Yeb,* a name which goes back to very early times (it would appear that the island of Yeb may anciently have been a place where ivory was bought from elephant-hunters who brought it down stream). 2. The hill upon which the Aga Khan's tomb is built overlooks the island which bears no resemblance to the shape of an elephant, a fact supported by maps. This does not, of course, make allowance for the probability that the shape of the island has almost certainly changed over the millenia; in which case von Däniken's pretended modern confirmation of his claim falls to the ground.

Similarly von Däniken claims that Egyptian mummification was an art imparted to the inhabitants of the Nile Valley by astronauts who knew how to restore the dead to life. Such a statement ignores the fact that sophistication in methods of mummification developed in Egypt over a long period, and it is possible to trace and set out how this took place, step by step. It began with observation of the simple fact that a body left in desert areas will desiccate and, unless destroyed by animals, will last almost indefinitely in desert areas — thus in Australia the desiccated bodies of persons lost in desert country have been known to survive for many years. The Egyptians aided this process by removing those parts of the body in which bacteria normally dwell and which, when death inhibits the production of protecting secretions by the body, attack their former home. The removal of viscera and brain can scarcely have been considered by von Däniken's highly scientific astronauts as conducive to revival of a dead body. He is likewise misled in his interpretations of the heavenly journeys made by the king in his personification of Horus — neither are they to be taken literally.

Von Däniken's book is superficially plausible but cannot stand up to examination. Apart from individual points of fact his whole claim is invalidated by the simple objection that if space contact with earth were deliberately organized why were its effects not concentrated in one locality where some enduring influence could be exerted rather than spread all over the earth's surface and often in such inconvenient places?

REFERENCES

1. BRIGHT, J. *Has Archaeology Found Evidence of the Flood?* in *The Biblical Archaeologist Reader*. N.Y., Anchor Books, 1961. p. 32 ff.
2. WOOLLEY, SIR L. *Ur of the Chaldees.* London, Penguin.

Memories of the Future

GEOFFREY STEPHENS
Chaplain, Perth College, W.A.

The Reverend Dr. G. H. Stephens, B.A. (Melb), M.A., D.Phil (Oxon), is a young Anglican priest, writer and scholar. He has been Chaplain at Melbourne Grammar School, and Keble College, Oxford. At present he is Chaplain to Perth College, Western Australia. He writes as a modern theologian.

How misleading the cover of a book can be! *Chariots of the Gods?* is variously subtitled 'Unsolved Mysteries of the Past' and 'Was God an Astronaut?'

In fact *Chariots of the Gods?* is not really a book exclusively about the unsolved mysteries of the past. Von Däniken uses the past and various mysteries which he considers unsolved or solved erroneously in order to urge a Utopian Archaeological Year in which scientists of all types will investigate the world and its past and utilize the clues and secrets which another race and genus left behind aeons ago. These clues and secrets were intended to be understood by man when he reached a certain level of sophistication. Von Däniken believes that properly understood they will give us a short cut to future development. So his concern is not so much the past for its own sake, but the future and its realization.

Likewise the other subtitle 'Was God an Astronaut?' is misleading. Von Däniken makes much of the idea that these primeval astronauts were mistakenly taken for gods by primitive man. He uses all kinds of evidence — giant stones, unique scientific advances, old cave and rock drawings, myths and folk legends — to prove his point. He himself, however, is no doubter. Primitive man was confused in his beliefs. The god-like visitors suffered from a case of mistaken identity and were merely visitors from another planet more sophisticated than ours. The question of the nature, existence and activity of God is untouched by von Däniken's theories. He says so himself quite explicitly (p. 70).

Neither 'Unsolved mysteries of the Past' nor 'Was God an Astonaut?' are very appropriate subtitles. Perhaps it would have been better to use the original German title *Erinnerungen an die Zukunft — Memories of the Future.* This is a very subtle and challenging title which describes the whole scheme of the book. Perhaps, however, it lacks the popular appeal of 'Was God an Astronaut?'

Was God an Astrounaut? Of course not, because God is not a thing, or an object, or even, strictly speaking, *a* being, but 'Being' itself — ultimate reality. Everything than von Däniken has written is quite remote from theology — it is merely interesting and exciting speculation about archaeology, and challenging and buoyant dreams about the future.

Von Däniken, however, does believe that his theories have theological implications. He thinks (p. 70f.) that theologians really do believe that what the Bible says about the creation of the world, the history of the Jews and the visions of the prophets, is literally and truly historical and that this is all contained in the dogmas of the Church and the pages of the Scriptures. Most theologians regard questions about the manner of creation as scientific problems in the first place. Most theologians would again argue that the Bible is not 'history' in the bald, rational sense. It is 'geschichte' (interpreted history) rather than 'historie' (chronicle); the spiritual history of a 'people of Hope'. The purpose of the Bible is religious, moral and mystical. The naive literal interpretation of von Däniken reveals no awareness of the value of the last 100 years of Biblical scholarship.

Most theologians agree with psychiatrists in describing some of the prophetic visions as para-normal psychological events which are visionary in character and which correspond to no reality. Again, nothing like the speculations that von Däniken makes could be built out of these personal and peculiar experiences of minds like Ezekiel, highly elated or deeply moved.

A Monopoly on God

Von Däniken seems to believe that the discovery of life (or even evidence of past lives) on other planets will shake the foundations of our religions and throw our past into confusion (p. 71). Of course there is at present no proof of other existences, but such proof would not alter for one moment the Christian belief that life is abundant and various, and that quite probably other forms and shapes sing praise to God on other planets. It is not as if Christians claimed to have a monopoly on God. There is, of course, only one God, but men on earth and even perhaps creatures on other planets praise him with their own tongues, within their own thought worlds, and under whatever form or forms they have inherited.

Von Däniken seems surprised at his own discovery that with the 'step into the universe we shall have to recognize that there are not two million gods, not twenty thousand sects or ten great religions, but only one.' (p. 71).

Theologians in the Judaeo-Christian tradition have known this for two millenia. Perhaps they have not always been tolerant of others, but any 'step into the universe' will not confuse theology but illuminate and invigorate it.

On Truth

Von Däniken also makes two theological statements about truth and history. He believes (p. 72f) that Christians claim an exclusive right to truth. In one sense Christians do claim to have the fulness of religious truth and at times they have acted cruelly towards those who claimed a similar monopoly. This is no longer the case. All truth is one, and by its nature it is 'true'. So if Christians have some truths it does not mean that other truths, be they scientific, historical or religious, are therefore excluded. Christians continually look for a deep enrichment of the truth which is at the centre of their faith — indeed truth is always a welcome visitor. The purpose and goal of life, as von Däniken says, is indeed to seek the truth (p. 73), but also it is to believe the truth once it is found. That is why it is quite reasonable to seek it under the aegis, although certainly not within the confines, of one's own religion.

On history, von Däniken has rightly discovered that 'everything, both myth and actual experience, makes up the history of a people.' (p. 73).

This insight into history has been the basis of much recent theological investigation and indeed there are whole schools of theologians who 'de-mythologize' the Bible and whole schools who attempt to 're-rationalize' the myths. Von Däniken's insight is the very first base that the young theologian must reach.

On Mankind's Purpose

Finally, von Däniken theologizes about man's ultimate goal: he believes that the whole human task consists in colonising the universe and that man's whole spiritual duty lies in perpetuating all his efforts and practical experience. Then the promise of the "gods" of peace on earth and that the way to heaven is open can come true.

The salvation of the human species from over-population, pollution and atomic destruction, and the alluring call to conquer space are grand and challenging demands in themselves. These would be indeed the means for a certain temporal salvation, but they in no way touch the personal and individual needs of contemporary men and women.

The triumph of love, the building of complete persons, the happiness of self-fulfilment — these are what religion and life are all about. The true salvation of man in the realization that here and now he can be complete and happy in love and life. The symbol of 'the resurrection of Jesus' indicates, for a Christian, that ultimately hatred and death are being, and can be, overcome by the forces of love and life. The symbol of 'God' indicates that it is Being itself, the life force, which originates and contains the power of love and life. The symbol of 'eternal life' indicates that what we experience *here and now* of love and life has a quality and a depth which is unshaken in the face of hatred and death.

Theologically *Chariots of the Gods?* is naive.

The 'Chariots' and the Dead Sea Scrolls

BARBARA THIERING

Barbara E. Thiering, B.A. (Sydney), B.D. (London), Theol.M. (Melbourne), part-time lecturer in Hebrew and Biblical Studies, University of Sydney. She is a well-known lecturer and writer. One of her special interests and areas of research are the Dead Sea Scrolls.

Von Däniken's statements about the Dead Sea Scrolls reveal ignorance of basic fact and basic methodology.

He includes among 'hitherto unknown texts'[1] belonging to 'finds during recent years on the Dead Sea (the Qumran texts)' an extract on an appearance of a chariot to Adam and Eve, from the *Apocalypse of Moses,* paragraph 33. The *Apocalypse of Moses* is not an 'hitherto unknown text'. It has been widely known since the fourteenth century A.D. It is also of no historical value, being typical of the highly speculative pseudepigraphic literature produced by Jewish writers in the early Christian period.[2]

The author appears to have realised his error while writing this paragraph, so he adds the saving words: 'This story, incidentally, does not tell us much that is new. Nevertheless, chariots of light, wheels and smoke were spoken of as magnificent apparitions as early as and in connexion with Adam and Eve.' Were spoken of as early as Adam and Eve!!! But if he simply means that they were connected with Adam and Eve, this is of no value until and unless the historicity of the whole literature, which *prima facie* is far from historical, is established on a number of independent grounds.

The Lamech Scroll, by which the author means *The Genesis Apocryphon* (the former title was used for a very short time, but abandoned as erroneous),[3] does contain, as he says, the story of the debate on the conception of Noah. The summary is accurate enough, but details have been added, without acknowledgment, from another writing from the period, *The Book of Enoch,*[4] which tells the same story. Both of the sources, the *Genesis Apocryphon* and *Enoch,* date from the 2nd-1st centuries B.C., being written by religious groups which indulged freely in speculation on ancient texts. As Millar Burrows remarks on the former,[5] 'Naturally this late version of the stories adds no authentic historical information to what is given in Genesis.' Von Däniken's word 'information' is hardly appropriate to this literature.[6]

There *is* a passage on chariots in the Scrolls, but surprisingly, von Däniken does not use it. Part of the

Angelic Liturgy,[7] it is a fanciful re-writing of the vision of chariots which Ezekiel records in the Biblical book of that name, ch. 1. This Biblical passage later became such a source of imaginative overgrowth that it was proscribed for liturgical use.[8] All descriptions of heavenly chariots in literature of the intertestamental and early Christian periods derive directly from this Ezekiel passage, being speculative re-workings of the same theme. They are therefore *not* independent evidence. There is one record only of an appearance of a chariot, the one that Ezekiel saw in 593 B.C. Most commentators treat this as belonging to a different order of sight than normal human vision, corresponding to the modern notion of symbolic or poetic vision.

No reputable historian would:

a. treat evidence as independent when it was not,
b. claim texts as new when they are not,
c. combine separate sources without acknowledgement, or
d. fail to deal adequately with accepted opinion contrary to his own about the nature of his sources.

Von Däniken, by his grave methodological errors, discredits all that he has to say about the Dead Sea Scrolls.

REFERENCES

1. P. 60, *Chariots of the Gods?*
2. CHARLES R. H. *Apocrypha and Pseudepigrapha of the Old Testament.* Vol. II. pp. 124, 129.
3. CROSS, F. M. *The Ancient Library of Qumran and Modern Biblical Studies,* Duckworth, 1958. p. 25, note 43.
4. CHARLES, *op. cit.* p. 278 ff.
5. BURROWS, M. *More Light on the Dead Sea Scrolls.* London, 1958. p. 8.
6. *Chariots.* p. 61, paragraph 3.
7. VERMES, G. *The Dead Sea Scrolls in English,* Pelican Books, p. 212.
8. DANBY, H. *The Mishnah.* Oxford, 1933. The references are: *Megillah* 4:10, *Hagigah* 2:1.

The Spaceman in the Tree

GORDON WHITTAKER

Born in England; emigrated to Australia in 1956. In 1961 he took up Roman Imperial History as a hobby, resulting in the discovery of a Roman fort in southern England in 1966. He changed his field of specialization from Roman to Aztec Civilization around this time, and began writing a thesis on loanwords in the Aztec language, which was completed in 1969. Through the auspices of Dr. Cyrus H. Gordon, the eminent Semitist and historian, he was given a Wien International Scholarship in 1970 to study at Brandeis University in the United States, where he is studying Mediterranean Civilization and its influence on the outside world, while independently researching the ancient history of Mexico. In 1972 he was appointed Lecturer in Aztec Civilization at Brandeis University.

It's very easy when reading *Chariots of the Gods?* and its sequel, *Return to the Stars,* to be led to the conclusion that our Big Brothers in Outer Space had a special affection for the lands of the Western Hemisphere, since von Däniken's works abound in references to the marvels of civilization they left behind there — far too advanced for mere Earthlings to have created.

Von Däniken takes advantage of the fact that the history

Maya Hieroglyphs, like these from a temple doorway at Palenque, were a sophisticated form of writing. A System of dots, bars, faces and hands enabled scribes to record dates and events.

of the civilizations in that hemisphere are neither common knowledge nor taught in high school in even the barest of outlines. Because of this lack he is able to distort the facts in any way he wishes, free from the fear that the general public, including students, will have the knowledge to see through his claims.

In an attempt to help compare the known facts of Western Hemisphere civilizations with von Däniken's statements about them, I will describe here those that I have been able to examine. Let us start with Ancient Mexico.

The Olmecs

Civilization began in Mexico around 1500 B.C. with the Olmecs, or 'People of the Rubber Lands', who built their religious centres in the mangrove forests of what is now Tabasco and Vera Cruz. They were an extremely artistic people, and worked wonders in stone. Although they appear to have lacked metal tools, they used stone implements with much skill to sculpt in basalt huge stone heads several feet high and weighing many tons. A dozen or so of these have been found at the sites of La Venta and San Lorenzo, and one famous example that often adorns books on early Mexican art is a head 9 ft tall from San Lorenzo.

Here is where von Däniken steps in (p. 117). Acknowledging that Olmec sculpture is amazing, he goes on to say that the stone heads are too heavy to move and so will never be displayed in museums. Bridges would not be able to support their weight. Well, not only are such giant heads to be found in museums, as is the 9 ft one I mentioned, but one was recently transported thousands of miles to the Metropolitan Museum of Art in New York for a special exhibition. Certainly they are heavy and difficult to move, and the Olmecs had to haul them some 46 to 62 miles to get them from the nearest quarries to the religious centres, but doubtless they considered, like other ancient peoples, the honouring and enrichment of their capitals well worth the physical effort.

The heads, which probably represent Olmec priest-kings since they vary considerably in appearance, were sculpted in basalt, a hard rock, so as to provide a more lasting monument to their civilization. This is fortunate, since in late Olmec times attempts were made to destroy them. Whoever was responsible eventually had to give up, and instead resorted to toppling them into ravines and burying them.

The Pyramid-City of Teotihuacan

With the disintegration of the Olmec Empire in the last centuries B.C., the seat of power and authority passed to the city of Teotihuacan in the central Valley of Mexico. In the film *Chariots of the Gods* von Däniken says that its age is shrouded in mystery. In the book he claims that all its structures are aligned with the stars, and that the most ancient text on Teotihuacan states that the gods assembled there to discuss mankind before Homo sapiens existed (p. 120).

Let me try to clarify the situation. The age of Teotihuacan is known fairly accurately through the combination of archaeological, historical and calendrical data. Its beginnings as a town date to about 150 B.C. and its foundation as a city occurred in the 1st century A.D., at which time pyramid-construction took place on a large scale.

Such vast structures as the often-photographed Pyramids of the Sun and Moon were erected on the interminable Street of the Dead, the main axis of Teotihuacan, which is aligned 15 degrees off north. This street, which is intersected by an east-west avenue, forms the basis for the grid of the entire city. Since all buildings are aligned with it, it would be wrong to say that they are aligned with the stars, as that is an argument that can only be applied to the axis itself. If the archaeologists who think the axis is astronomical in nature are right, it will not be an earth-shattering discovery, as it is a relatively simple matter to line up two points with a star or planet. All it takes is a good eye and two sticks.

And as for the oldest text on Teotihuacan, it is in fact quite recent — the end of the 16th century A.D., several decades after the Spanish Conquest of Mexico. It is written in the Nahuatl language of the Aztecs and says absolutely nothing about Homo sapiens. What it does say is that the gods got together at Teotihuacan to decide which one of them should sacrifice himself in order to become the sun. The sacrifice required the sore-infested god Nanahuatzin, whose name means 'Most Venerable Mushroom', to hurl himself into a great fire. It is difficult to imagine what this could possibly have to do with astronauts. And the Aztec text refrains from mentioning the anthropological classification 'Homo sapiens', which was only coined in modern times.

Tollan and Chichen Itza

After the fall of Teotihuacan in the middle of the 7th century A.D., Mexico lacked a dynamic centre of civilization until the rise of the city of Tollan, the modern Tula in the State of Hidalgo, in the last half of the 10th century A.D. Its people, known as the Toltecs, blended new and old ideas to create a vibrant art-style which was equally at home in Tollan as it was in the conquered Maya city-state of Chichen Itza, several hundred miles away.

Von Däniken had a field day with Toltec civilization in his film and books. In the film he shows the Pyramid of Tlahuizcalpantecuhtli, which is in the Toltec capital, and devotes special attention to the tall statues on top. To von Däniken they are helmeted spacemen wearing box-like units on their chests and holding sophisticated weapons or communications equipment in their right hands. It does not take a close examination to notice that the helmets are nothing more than feathered head-dresses. The units on their chests are protective breastplates worn by Toltec soldiers, and the communications equipment are only spear-throwers, known as atlatls to American anthropologists and as woomeras to Australians.

Chichen Itza figures prominently in von Däniken's works. He emphasizes that all Maya structures were built according to the dictates of the Maya calendar (p. 120), and to illustrate his claim he states that the Toltec pyramid at Chichen Itza known today as El Castillo — The Castle, has the same number of steps up it as there are days in the year. It takes little more than patience to work out how many days there are in a year — you note where the sun sets on a certain day and count how many days it takes to reappear at that spot. Astronauts are not needed for that.

The observatory at Chichen Itza crops up both in the film and in the book *Chariots of the Gods?* (p. 125). The film tries to pass on the impression that the observatory is dome-shaped by filming it from an angle where the slope of the ruined wall resembles the slope of a dome. He enhances the effect by comparing it with the Mt. Palomar Observatory. In actual fact, although the building is circular, its walls are straight and its roof flat. The roof has long since fallen through, and part of the wall has crumbled to its modern deceptive state.

The observatory has very narrow slits in its walls, through which for a few seconds the sun's rays pass on only two occasions in the year. This enabled Maya astronomers to calculate time with great precision. In all probability the observatory was not intended for use in scanning the heavens.

Von Däniken then gives a short account of the sacred well, or cenote, of Chichen Itza (p. 126-7), after which he makes the startling revelation that in fact there exists a second well, identical to the first, which is never mentioned by modern scholars because it does not fit into our theories. Von Däniken strikes again, bringing another closely guarded secret to light — or so he would like you to believe. Actually, the two wells figure on all the maps of the city. They are not by a long shot identical, nor are they perfectly round, and whereas the sacred cenote had an obvious religious function, the other supplied fresh water for the city's population.

Von Däniken is also confused about the origin of the wells and believes they were caused by meteorites crashing to earth or by rockets blasting off. The word 'cenote' is derived from the Maya *dz'onot* and refers to the natural wells found all over Yucatan. These wells, which provided the primary source of water for the Maya, were formed when the limestone land-surface collapsed over underground water.

In *Return to the Stars* there is a line-drawing which von Däniken describes as showing Maya priests wrenching the hearts from live victims, perhaps, he thinks, in memory of a misunderstood surgical technique employed by our spaceman visitors.

The line-drawing is from a gold disk recovered from the sacred cenote of Chichen Itza. It shows Toltec, not Maya, priests in characteristic dress performing a human sacrifice to the sun god Tonatiuh. According to Toltec belief, the sun needed human hearts to sustain it and give it energy to continue on its course through the heavens.

Without them the sun would die and the world would come to an end.

The Aztecs and the Prophecy of Quetzalcoatl

Toltec civilization crumbled with the fall of Tollan to northern barbarians in 1168. After more than a century of anarchy in Central Mexico, the Aztecs founded their capital, Mexico Tenochtitlan, on an island in the middle of a lake, and soon after embarked on a series of conquests which eventually stretched from Acapulco to Nicaragua. After only a century of existence, the Aztec Empire was brutally destroyed by the Spanish under Hernan Cortez.

Contrary to von Däniken's statement in *Chariots of the Gods?* that Cortez had a devastating impact on South America, it must be pointed out that South America begins at Tierra del Fuego in the south and ends at Panama in the north, and that Mexico, far from being in South America, is half-way up the North American continent.

Von Däniken appears to have gone out of his way to distort and ignore the known facts of Aztec civilization. For example, the Aztecs had two kinds of drums, the huehuetl or upright drum, and the teponaztli or horizontal slit drum. In the film *Chariots of the Gods?* von Däniken calls a huehuetl a space vehicle, and in *Return to the Stars* he describes a teponaztli as a ceremonial vessel depicting a circular space-ship (actually a solar disk).

Among the photographs in *Return to the Stars* there is one of the Aztec calendar stone, here labelled as a round Maya calendar, although the two civilizations are quite distinct and there is no way the calendar stone can be considered Maya. Nor are telescopes necessary items for constructing a calendar, although von Däniken obviously thinks otherwise.

Von Däniken often gets himself into difficulties confusing one civilization with another until they are both nearly unrecognizable. Take for instance the Plumed Serpent deity of the Ancient Mexicans. In the Aztec language, Nahuatl, he is known as Quetzalcoatl. He was accepted by the Yucatan Maya under the name Kukulcan, and by the Quiche Maya as Gukumatz. A Toltec ruler who was born on the day sacred to Quetzalcoatl, a god of the planet Venus, took the god's name as a title, causing him to be confused with the god.

Yaxchilan. A typical type of the Maya ideal of beauty—artificially deformed head with sloping forehead, almond-shaped eyes, large nose, drooping lower lip and receding chin.

In *Chariots of the Gods?* (pp. 127-8) von Däniken says that the Maya god Kukulcan is probably the same as the later god Quetzalcoatl, and that there are Maya legends about Quetzalcoatl. Unfortunately, he has the facts upside down. The Toltec Quetzalcoatl is earlier than the origin of the Maya Kukulcan, and accounts of Quetzalcoatl are Toltec, not Maya. The Maya describe only Kukulcan.

And as for the legend, von Däniken outlines it as follows (p. 128). The bearded, white-robed Quetzalcoatl came from a land to the east, handed down laws and civilization to the inhabitants, then took a ship to the planet Venus after promising to return. He is also credited with growing man-size corn-cobs and putting the colour into cotton.

The god and the man have been confused here. The god (and I do mean god — not astronaut) is associated with Venus and is the deity of agriculture and fertility, whence the corn and the cotton. The man, or more likely last of a series of men with the name Quetzalcoatl, was the son of an adventurer who crossed the sea to the east, presumably the Atlantic, and conquered Mexico in the late 10th century A.D. This son was made King of Tollan by his father, and ruled the city peacefully for two decades until the unpopularity of his religious reforms caused a civil war which resulted in his overthrow. He fled to the Gulf coast swearing that one day, on the anniversary of his departure, people from the eastern lands would return and reclaim Mexico. The prophecy came true to the letter when, in 1519, the Spanish flag appeared on the Aztec coast.

I feel there is sufficient evidence to suggest ancient contact with the Americas over the Atlantic and Pacific. At present, though, scholars are divided on the issue, and in any case it is best to approach the subject and the evidence with extreme caution.

Von Däniken claims that Quetzalcoatl can not have been from the Old World because the first things he would have given the Maya would have been the cart and the wheel, and, as he says, they used neither. However, that is still the case, although the wheel is well known throughout Yucatan and Guatemala. The Maya have no use for the wheel. In one religious festival in Chichicastenango, Guatemala, the Maya actually carry a car in which is a statue of the Virgin Mary. In ancient times a cart would have been useless since there were no draught animals to pull it — no horse, no mule, no burro, no bullock.

Easter Island

After reading through von Däniken's chapter on Easter Island in *Chariots of the Gods?* (pp. 113-9), the reader is supposed to come away with the distinct impression that in the dim past a mysterious unknown people carved all the mysterious giant statues there in one go, using other than conventional stone tools, and mysteriously transported them to the coast of the island, using other than conventional wooden rollers.

Apparently, von Däniken prefers to ignore the fact that

the Easter Islanders were Polynesians, as are their living descendants; that they probably came together with their culture from the nearby Marquesas Islands within the last 1500 years; that they carved over a long period of time several hundred statues most weighing under 50 tons, using the stone tools that have been found there in large numbers; and that their own native accounts say that the statues were hauled about with thick ropes and stone, not wooden, rollers. The statues resemble religious monuments

The method of erection of Easter Island figures as suggested by Thor Heyerdahl.

found on other Polynesian islands. It is probable that the cyclopean masonry of Easter Island is due to Peruvian influence, since it resembles South American architecture. Inca historical accounts mention the exploration of the Galapagos Islands in the 15th century by the fleet of an Inca emperor, so there may well have been other such voyages in earlier times. Again, there is no need or evidence for astronauts.

Nazca International Airport

Now that I have mentioned Peru, I might as well discuss von Däniken's Nazca Airport. In the film he states quite dogmatically that there is no doubt that the markings are the landing fields of an airport. It is not quite that simple.

Although the lines and animal patterns can only be seen from the air, many of them stretch for miles in every direction. Aircraft using them would have to have had impossibly poor brakes or the equivalent if they had to run along the 'tarmac' for such distances. Besides, astronauts would have landed in space-ships, not aeroplanes, and these come down vertically.

Archaeologists freely admit that they do not know the reason for the markings, but their theories to date are sounder than von Däniken's. The lines appear to be astronomical and calendrical in nature, since many line up with stars and others become visible when constellations rise. The animal patterns may represent the constellations. With a knowledge of surveying, these would have been no problem to lay out and etch into the desert surface. In *Chariots of the Gods?* von Däniken proposes that the Peruvians lengthened the lines that the astronauts had marked, but in that case the extensions would surely have not been as accurate as the originals.

There is no easy solution to the problem of the Nazca lines at present, but at least it is possible to see which theories do *not* fit the facts.

A 'With-it' Astronaut and his Wooden Rocket

It would only be fitting to end the discussion of an imaginary airfield with that of an imaginary rocket. On the cover of the hard-bound *Chariots of the Gods?*, in the text (pp. 123-4), in the film, and on the dust-jacket of the record, a rocket in Mayan art is on prominent display. Pooling all the descriptions von Däniken gives of it, one gets the following description of the A.D. 692 Maya sarcophagus relief from the Temple of the Inscriptions at Palenque in southern Mexico.

An astronaut is at the controls of his rocket. He is at the height of fashion in his shorts, an internationally tailored jacket, a tube up his nose and an antenna in his hair. Leaning forward a little, he is fiddling with the controls, pressing his foot down on a pedal, and staring at some instruments near his face. The rocket is at lift-off.

However — and this might distress the Gould League — there is a casualty. It's a rather absent-minded quetzal bird (the national bird of Guatemala) that is perched on the nose-cone of the rocket at the top of the picture. Von Däniken does not mention him. But then, he does not mention the rather distasteful skull with the big fangs pictured below the astronaut's seat either. Nor does he mention the life-expectancy of the astronaut, who is sticking his head outside the rocket.

And now that I think about it, the astronaut must have taken his trendy jacket off, since all I can see is a necklace and ornament on his chest. His shorts are sadly too much like the commonplace Maya kilt — he probably picked it up on his travels. His antenna could easily be mistaken for a Maya hairdo — camouflage, I suspect.

One has to hand it to von Däniken for not being fooled by the arrogant members of the Establishment who say that the relief only shows a Maya dignitary on a throne above the skull of the Lord of the Earth, and that the dignitary is about to pluck a fruit from the Tree of Life, on top of which is a sacred bird representing the zenith direction.

After all, who ever heard of an astronaut in the company of a bird blasting off into space in a tree?

BIBLIOGRAPHY

LEONARD, J. N. *Ancient America.* Time-Life, 1968.

ANTON, F. and DOCKSTADER, F. J. *Pre-Columbian Art.* Abrams, 1968.

SOUSTELLE, J. *Mexico.* World Publishing Company, 1967.

NICHOLSON, I. *Mexican and Central American Mythology,* Paul Hamlyn, 1967.

THOMPSON, J. E. S. *Maya Civilization.* Univ. of Oklahoma, 1966.

BAUMANN, H. *Gold and Gods of Peru.* Oxford Univ. Press, 1963. and the article *Easter Island* in Collier's Encyclopedia.

Ancient Egypt Without Astronauts

CHRISTOPHER BAKER
Professor at St. Columban's College, Sydney.

The Reverend Father C. J. Baker, D.D., L.S.S., is a professor at St. Columban's College, Sydney. He finished his studies at The Pontifical Biblical Institute in Rome in 1955 and has since lectured at Sydney University. He is a student of ancient Egyptian language and culture, and because it brought back happy memories of his earlier visits there, he enjoyed the film. But not the script.

My introduction to *Chariots of the Gods?* was through the film. The fine coloured shots of marvels of antiquity (and more recent times) renewed something of the pleasant wonder with which I had gazed on the works themselves in such places as Thebes and Luxor, Saqqara and Gizeh, Heliopolis and Cairo, Babylon and Ur of the Chaldees, Nineveh and Baghdad, as well as the more recent structures in Cuzco, Sacsahuaman, Machu Picchu, Tiahuanaco and Easter Island. The movie script, however, detracted considerably from the value of the production, because it kept attributing man's greatest achievements in antiquity to the influence of visitors from outer space. Added to this was the assumption that the religion of those times arose from a mistaken idea that the astonishing visitors from other planets or stars had been gods, not just highly intelligent astronauts dropping in for a friendly visit. On reading the book on which the film was based, E. von Däniken's *Chariots of the Gods?,* I found that the book was still more insistent than the film with regard to the main hypothesis. As the author keeps asking what he seems to regard as unanswerable questions for those who would reject his theory, it is only fair to offer him (and any puzzled reader) some answers concerning the marvels of ancient Egypt.

Von Däniken's treatment of Egyptian antiquities has to be seen in the wider context of his dominant hypothesis about visitors having landed on earth from outer space and exercised a decisive influence on the development of mankind. It is not easy to isolate the basic details of his hypothesis with regard to the following three points:—

1. The time of the supposed astronauts visits to earth: He concedes the reliability of history for 2,000 years back, being concerned only with 'the most remote antiquity (p. 111). This does not stop him dealing with the Inca civilisation around Peru and Bolivia, where it flourished from about 1000 to 1500 A.D. (cf. J. A. Mason, 'The Ancient Civilizations of Peru', p. 104). He suggests that 'unknown intelligences from the universe' began to influence earthlings 'in the Early Palaeolithic Age, i.e., between 10,000 and

Pyramid of Medium. *(see p.67)*

A quarry gang at work at Aswan. Tipping a block and easing it on to rollers. *(see p.68-p.70)*

Part of the design on sarcophagus lid, Temple of the Inscriptions, Palenque, probably A.D. 692. *(see p.20, p.60)*

The Maize God (San José, British Honduras.) A typical animated Mayan form of the late Classic Period 680-800 A.D., displaying strong similarity to the figure at Palenque. *(see p.56)*

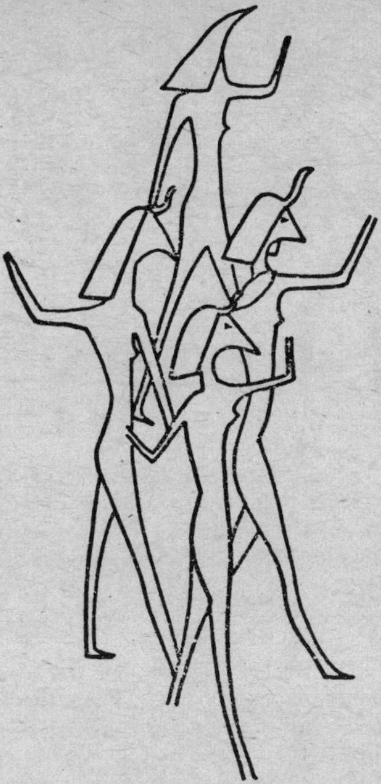

Jabbaren. Bird headed Goddesses from the Tassili Frescoes in the Sahara. *(see p.20, p.33, p.62)*

Stone Pyramids of Dahshur, viewed from the north-west. *(see p.67)*

Jabbaren. The Judges' Post Bovidian Period with Egyptian influence. Rock paintings of the Sahara. *(see p.62)*

Sefar the Great God with praying women. Decadent Phase of the Round Heads, cave drawings in Sahara. From the Tassili Frescoes. *(see p.20, p.53, p.62)*

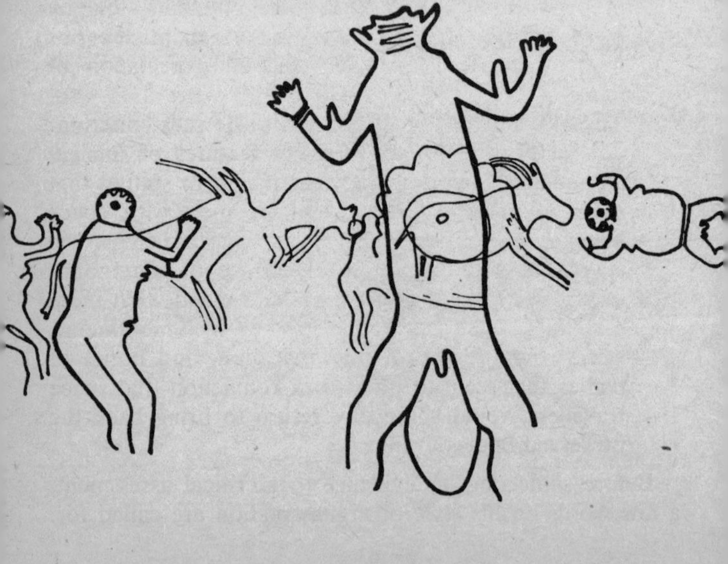

40,000 B.C.' (p. 111-2). (Given the controversies over the dating of the Stone Age, suffice to note here that archaelogists like W. F. Albright and R. North consider 10,000 B.C. as very late Stone Age or Palaeolithic.) A similar dating is implied when von Däniken asserts that the space visitors could have made predictions for our planet '10,000 years ago' (p. 179). In another passage he really goes well back into Stone Age days when he posits that highly developed civilisations, superior in technology to our own, 'may have existed hundreds of thousands of years ago' (p. 167).

2. The number and purpose of the supposed visits: von Däniken postulates that there were at least two visits by the same kind of space beings, who on one visit artifically fertilized women, then 'thousands of years later' returned to continue their breeding experiments. From these came sufficiently intelligent humans to found communities, to paint wall caves, to make pottery, etc. On the other hand, the space visitors exterminated the poorer specimens or took them off to other continents (pp. 71-2). The astronauts are presumed to have been keen to help humans along by these means, also by imparting to them superior wisdom, better techniques, and an expectation of rebirth.
3. The evidence of the alleged visits: He sees confirmation of his hypothesis in many features of ancient monuments, e.g., drawings, bas-reliefs, or statues that seem to represent spaceships or men with space helmets, etc.; gods depicted as winged beings; the magnitude and incredible excellence of the monuments which seem to demand far beyond what mere earthlings could have achieved when such monuments were made; forms of mummification, and burial of bodies that seem to point to a conviction that space travellers would some day return to bring back the preserved bodies to life.

Before subjecting his evidence to a critical assessment, a few words on his style of argumentation are called for.

In general one notices throughout his book a tendency to slip from establishing that a thing is *possible* to an assumption that it *really exists* and so can be used as a 'proof' of his own hypothesis (cf. p. 20; p. 106; p. 167). He often encourages the reader to make the jump for himself by a loaded rhetorical question. At the same time he attempts to forestall would-be critics of his theories by calling into question their competence, their openness to new ideas, their objectivity (cf. p. 11; p. 107; p. 124). It is my contention in what follows that a more extensive, careful and objective reading of archaeologists, historians, and other experts on ancient civilizations would supply much more convincing answers to the riddles of Egypt's past than the hypothesis put forward by von Däniken. It must inevitably show up some of the weaknesses in his theory.

Egyptian civilisation did not suddenly drop to the Nile banks from some star. There is plenty of evidence of settled village life around Fayum in the north from c. 4200 B.C. and around Tasa in the south c. 4000 B.C. (cf. A. Gardiner, *Egypt of the Pharaohs,* p. 387 ff; J. Bright, *A History of Israel*, pp. 25-6). This allows a thousand years of development prior to the First Dynasty. During this period the Egyptian villages were organised for draining the swamps, irrigating their own crops, fruit trees and vegetables. They also learnt how to grow flax to provide them with linen and rope. Pottery was being made, and there is evidence (e.g. in cylinder seals and pottery types) of interchange of products with other countries.

Egyptian Writing

Egyptian hieroglyphic writing is attested right from the start of the First Dynasty (c. 3100 B.C.). Thanks to their ability to record transactions and events, often chiselling an account into granite, we are in a great position to reconstruct the salient points of their history from that time on. One of the questions raised by von Däniken is, 'who gave them . . . a ready-made writing?' (p. 86). Perhaps it would not appear so amazing to him if he did not think that the Sumerians only began to write c. 2000 B.C. (p. 40). Actually the Sumerians invented writing somewhere c. 3300-3100 B.C. of which texts have been found at Warka (= Erech) (cf. S. N. Kramer: *History Begins at Sumer,* p. xix; J. Bright, op. cit., pp. 23-4). As von Däniken notes (p. 69), the Sumerians had some influence on Egypt, and others maintain that the great impetus given to Egypt as the First Dynasty emerged was due to Mesopotamians arriving on the scene (cf. W. B. Emery, *Archaic Egypt,* pp. 38-42). However much their arrival may have sparked off interest in writing, the Egyptian language and its written form differed considerably from Sumerian. It does have some affinity with Semitic (incl. Babylonian and Hebrew) and Hamitic (East African) languages. Its hieroglyphic writing is a development from simple pictorial representation of an object. It became a combination of ideograms (sense-signs or pictures) and phongrams (sound signs or consonants). The neat hieroglyphics present an intriguing picture parade of ancient Egypt with its gods and temples, kings, palaces, work and home life, implements and clothing. This is important evidence that the language really grew out of the daily life and culture of the Egyptians themselves. It was not superimposed on them by any visitor from space, nor even from Mesopotamia.

Building a Pyramid

The Great Pyramid of Cheops (also called Khufu) at Giza near modern Cairo has been for over 4,500 years one of the greatest wonders of the world. Naturally von Däniken devotes many pages to it in a chapter entitled: 'Ancient Marvels or Space Travel Centres?'. He rejects all the explanations commonly advanced these days by the archaeologists and other scholars, on the grounds that they leave a host of burning questions unanswered. He offers alternative answers to some of his own questions, but he shrewdly avoids any details as to how its building would have been accomplished by space travellers. He implies, of course, that it would have been simple for superior intelligences. By his reckoning it would have taken the workmen of Cheops' era 664 years to build that pyramid! As it could not have been built in one king's lifetime, the inscription crediting Cheops with its construction must be just a forgery or usurpation on his part. The pyramid then, 'if that *was* the case', had been built long before the time of Cheops (p. 101).

We must answer to the principal questions or objections raised by von Däniken against the accepted position that the Great Pyramid was built in the time of the king Cheops (c. 2600 B.C.) of the Fourth Dynasty, using only normal human resources. Let us now run through the facts so painstakingly unearthed and co-ordinated by the Egyptologists over the past 150 years (i.e. since moderns learnt how to translate hieroglyphic writing and to unearth much more of it around Egypt).

1. *A gradual development of the pyramid:* Right from the start of the First Dynasty (c. 3100 B.C.) the kings were buried in a grave over which a rectangular, flat-roofed structure of mud brick was erected. This structure was a 'mastaba' and was sufficiently like a house to suggest that it was intended as such for the benefit of the buried king. Well preserved 'mastabas' have been found at Saqqara (cf. Edwards, *The Pyramids of Egypt*, pp. 37-40).

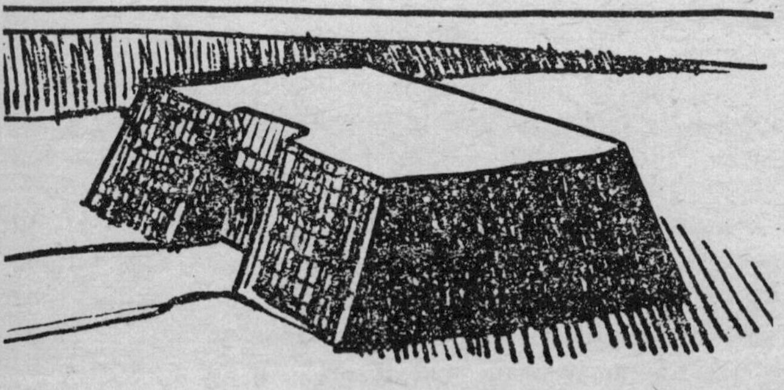

Flat topped Mastaba.

Next came the 'Step Pyramid' of Zoser (c. 2700 B.C.), a king of the Third Dynasty. His architect, Imhotep, is remembered as the one who initiated building in stone

©Lorillard 1973
KENT
WITH THE FAMOUS MICRONITE FILTER
DELUXE LENGTH
Micronite filter.
Mild, smooth taste.
America's quality
cigarette.
Kent.
King Size or Deluxe 100's.
Kings: 16 mg. "tar," 1.0 mg. nicotine; 100's: 19 mg. "tar," 1.2 mg. nicotine;
Menthol: 18 mg. "tar," 1.3 mg. nicotine; av. per cigarette, FTC Report Sept. '73.

KENT
MENTHOL
WITH THE FAMOUS MICRONITE FILTER
DELUXE LENGTH
KENT
MENTHOL
DELUXE LENGTH
Try the crisp, clean taste of Kent Menthol.
The only Menthol with the famous Micronite filter.
Warning: The Surgeon General Has Determined That Cigarette Smoking Is Dangerous to Your Health

instead of brick. At first the superstructure over Zoser's grave was flat topped 'mastaba' with a square base instead of the usual rectangular one. The mastaba was extended to make it rectangular, then a four-stepped pyramid was built incorporating the mastaba. Still another extension to the base was made and the pyramid rose to a six-stepped one (cf. Edwards, op. cit, p. 56).

Further refinements were introduced by the building of the earliest known true pyramid, i.e. one without stepped sides. This is the pyramid of Meidum, erected early in the Fourth Dynasty and thought to have been for king Seneferu c. 2650 B.C.). The partly ruined state of this pyramid makes it easier to discern that it was first built as a six-stepped, then a seven-stepped pyramid, after which the four sloping sides were completely smoothed and faced with limestone. Only then was it a true pyramid. Its changes of design are typical enough of normal human progress and learning through action; it does not demand a superhuman intellect as architect.

The same king, Seneferu, is named at least twice on inscriptions of the Bent Pyramid at Dahshur, which started off to be a true pyramid, then about half way up the angle of incline is reduced by about 11 degrees, as if to finish off the work quickly. It had been started on a 620 ft. square base and rose more than 300 ft.

At last the stage was set to undertake the Great Pyramid of Cheops/Khufu (c. 2600 B.C.), a king of the Fourth Dynasty who succeeded Seneferu. It is the biggest of all the Egyptian pyramids, having an *almost* square base of c. 755.5 ft. ('the differences between the longest and the shortest side was only 7.9 inches' — Edwards, op. cit., p. 118), and rising to 481.4 feet. It was originally faced with limestone, most of which has long since been removed. Some 2,300,000 stones, averaging about 2½ tons each, went into the pyramid. The name of Cheops is found painted on some of the blocks within the pyramid (cf. Edwards, op. cit., p. 127). The burial chamber for the king was fashioned deep inside the pyramid, and his empty sarcophagus is still within it — since it was cleverly made a fraction wider than the passage now leading to the chamber and must have been placed there before the pyramid was completed. Despite von Däniken's impatience with 'rehashed erudition' about the purpose of the pyramid being just a burial place for a Pharaoh (p. 98), it is obvious that the king himself took that aspect of it very seriously.

2. *The ability to cut stone, even granite:* In the Cairo Museum can be seen copper implements from the First Dynasty. Included among them are a copper saw and chisels. In a quarry near Aswan stands a large block of granite showing wedgemarks to facilitate splitting of the block from the rock face. Presumably wooden wedges were to be driven in and perhaps wet to make them expand. Around monumental tombs of the First Dynasty at Saqqara and nearby Helwan were some sawn stones weighing several tons, yet this was several centuries prior to the time of Cheops (cf. Edwards, op. cit., p. 160; Emery, op. cit., p. 226 & Fig. 128). Centuries of experience in quarrying stone can explain the production of stones in much greater numbers and in larger sizes. Even if we do not know everything about the early techniques of hewing stone, we should not simply conclude that the Egyptians could not have done it somehow by their own ingenuity.

3. *The manpower to construct the Great Pyramids:* The presumption is that many tens of thousands of men were employed over a period of about 30 years to build the Great Pyramid. Von Däniken objects that Egypt at the time could not have supported sufficient labourers to build it so quickly (hence he calculates a building time of 664 years). He argues that Egypt did not have enough food for anywhere near the number of labourers needed to build it in one king's lifetime. In reply, we point to the great fertility of the Nile Valley (and not just around Giza, because Cheops fully controlled 'Upper and Lower Egypt'). As noted earlier, even pre-Dynastic Egypt grew crops, fruit, vegetables, had herds of cattle, and caught plenty of game and fish along the Nile. The annual flooding of the Nile added top dressing to their fields, and the hot sun encouraged prolific growth where moisture was

also available (thanks either to flooding or irrigation). The Egypt which could store up plenty of grain to withstand a seven-year famine in the days of Joseph — a thousand years later — was obviously able to cope with food requirements in the days of Cheops. In passing, be it noted that Genesis cc. 37-50, which tell of Joseph in Egypt, enshrine a wealth of extremely accurate detail about 16th century Egypt, whereas Genesis ch. 6, which mentions 'sons of God' marrying 'daughters of men', deals with prehistoric times. Von Däniken (p. 52) argues, mistakenly I hold, from Genesis ch. 6 as though it were a normal historical record. He should not disregard the much later Joseph story.

4. *The equipment to move and place huge stones:* Presuming that Cheops had many tens of thousands of well fed workmen, was it possible for them to haul the millions of huge stone blocks into positions up to 481 ft. above ground level? While we admit that a tremendous amount of sheer physical energy must have been spent during several decades of construction, nevertheless there were several simple aids available, e.g. rope, wooden rollers, sleds, levers, boats, and earthen ramps. But what of von Däniken's taunting assumptions that such ropes and rollers, like the grain supply, were 'non-existent' (p. 101)? Let the real facts, not presumptions, decide the issue.

One of the important by-products of the linen industry which was already flourishing in pre-dynastic Egypt (i.e. before 3100 B.C.) was rope made from flax. One of the best examples of such rope was found in a tomb of the First Dynasty at Saqqara, hence dating from 3000 B.C. It is worth noting that well made leather thongs were also found in Saqqara tombs of that period (cf. 'Egypt' in *Encyclop. Britannica,* p. 47). Rope was also made from fibre, especially the heavier rope for hauling purposes (cf. Emery, op. cit., p. 223). Knowing the remarkable advances made in the design of burial monuments during those 500 years prior to Cheops, we are entitled to postulate that the production of rope likewise improved notably during that time. There would have been plenty of suitably strong rope to haul any stone used in the Great Pyramid. (Incidentally, the average weight of the stones used was only 2½ tons, not 12 tons as stated by von Däniken (p. 101-2). This factor alone should cut down his estimated 664 years by about 526 years!

To facilitate moving the building blocks from the quarries to the Pyramid, workmen certainly could have used, and presumably did use, some form of wooden rollers or sleds. What of the objection that Egypt had no wood apart from soft date-palm trunks (and the dates were too vital as food in any case)? Admittedly Egypt in those days had to import wood from other countries, but there is ample evidence that wood was in fact imported in large quantities — it can still be seen in extant monuments and boats from as early as the First Dynasty. The carpenters and boat builders obviously had plenty of cedar and cypress on hand, obtained from Lebanon and Syria (cf. Emery, op. cit., pp. 203-4, also pp. 185-8 for vivid drawings of early timber ceilings; cf. also Edwards, photo 9, for big beams within Bent Pyramid). It could have been shipped from Byblos right to the nearest Nile bank, and definitely not to 'Alexandria' as von Däniken thinks would have been necessary (p. 97). As its name suggests, Alexandria was founded by Alexander the Great, c. 331 B.C.

That the stones were sometimes dragged on wooden sledges is likely enough, especially when they weighed 15 tons or more. There is an Egyptian hieroglyph for 'sledge' from about that time. It includes an ideogram depicting a sledge with the front runners curving up and capped by a jackal's head. This word occurs in Pyramid Texts, so called because they were found on walls inside the pyramids of kings of the Fifth and Sixth Dynasties buried at Saqqara. Remember that Cheops belonged to the Fourth Dynasty. What is more, Alan Gardiner points out: 'These texts, for the most part *of very great antiquity,* are exclusively concerned with the welfare of the dead king . . .' (*Egyptian Grammar,* Oxford University Press, 1950, p. 18). It seems to me that the Narmer Palette (c. 3100 B.C.) depicts a sled with its runners terminating in a captive's head instead of the later jackal's head (cf. reproduction in Gardiner's Grammar, p. 7, or P. Montet, *Eternal Egypt,* p.44). There is a vivid picture of a colossal statue, estimated to weigh about 60 tons, being hauled on a wooden sledge by 172 men. They have 4 long ropes attached to the sledge, while the Colossus itself is tightly roped onto the sledge. One workman is busily pouring from a pitcher in front of the sledge, no doubt to cut down

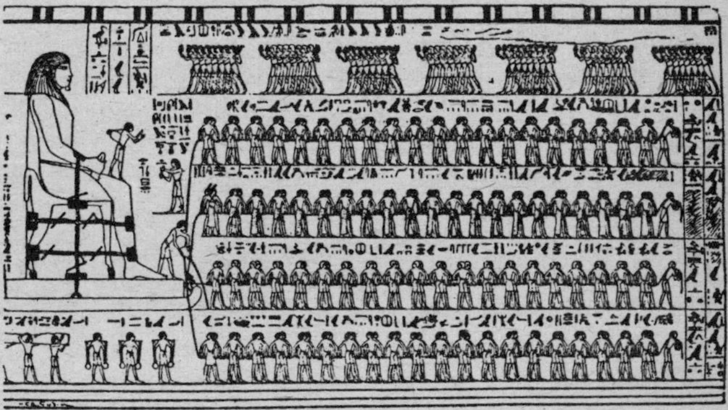

The transport of the colossus of Djehutihotep, a nobleman of the XIIth Dynasty, mounted on a sledge.

friction and heat. The drawing was found in the tomb of a nobleman, Djehutihotep, from the Twelfth Dynasty, (c. 1990-1790 B.C.). It was from the Twelfth Dynasty onwards that the large obelisks of 50 ft. to 100 ft. and the most grandiose colossi (e.g. of Rameses II in 13th cent. B.C.) were erected.

Wooden boats were also commonly used for transporting anything heavy, and the largest buildings, including the Great Pyramid, were erected within flooding distance of the Nile. People who had been irrigating fields for at least 500 years were well able to run channels close to the site of Cheops' Pyramid. That there existed at that time wooden boats sufficiently large to carry big stones is demonstrated by the discovery in 1954 A.D. of a wooden boat still in position in one of the five boat pits surrounding the temple complex connected with the Pyramids. The boat measures 143 ft. in length, and presumably was intended for the king's use as he accompanied the Sun-god on his daily journeys through the underworld by night and through the sky by day (cf. Edwards, op. cit., p. 133 and Photo 13a). This transport is by no means depicted as a modern space ship. Somewhat similar boat pits have been found at the tombs (mastabas) of early First Dynasty kings at Saqqara. The pits would hold a boat of about 50 ft. length (cf. Emery, op. cit., pp. 130-133). Well before that, perhaps early in the fourth millenium B.C., the Gerzeans in Lower Egypt were already able to build sizable boats with many oars and two cabins, as can be seen on their distinctive pottery (cf. P. Montet, op. cit., p. 41; A. Gardiner, *Egypt of the Pharaohs,* p. 391). In other words, by the time of Cheops the Egyptians had well over a thousand years experience in boat building and boating on the Nile. If they needed a boat for a particular purpose, they could construct and sail it.

Earthen ramps, sometimes strengthened by masonry, are still to be seen around Thebes and Karnak, but these are a thousand years or more later than the Great Pyramid. However, there are remains of a 400 metre causeway constructed to connect king Chefren's Mortuary Temple with a Valley Building, all part of the complex for the Second Pyramid built beside that of Cheops.

Von Däniken makes a great deal of his assumption that bodies of the Pharaohs were embalmed to await a restoration to life in the distant future when some space visitor returned to earth. As all their religious texts proclaim, the dead Pharaohs were really expecting their life to continue uninterrupted. The body was embalmed to ensure that the 'ka', the spirit or life-force or personality or double of the dead king, would still have as close to the same body as possible for continuing life.

As for von Däniken's theory that the many representations of winged gods in Egypt were a reflection of earlier visits from space beings, it is unfounded and unnecessary. The wings are obviously birds' wings, not rocket-craft wings. Much of the prominence of wings is due to the syncretism of the Egyptian religion, which allowed them to combine features of various gods like Amon, Ra, Aten, and Horus whose symbol was the Falcon. Other gods like Nekhbet were depicted as a vulture, Toth as an ibis, and the goddess Ma'at as a feather. Other gods were symbolised by animals and reptiles, an indication that their religious symbolism was developed by close contact with the nature-world of mother earth rather than with the outer stars. It is true that they had the Sun-god as their chief god, but that was due to his obvious warmth, life and power beamed at humans in a country that hardly knew what rain was.

Horus, the Hawk-God, from the palette of Narmer.

Space is insufficient now to take up the features that set apart the religion of Moses from pagan nature religions, for instance in its monotheism, its explicit forbidding of any image to represent God, its morality inseparable from its worship, its unique conviction that

God had drawn Israel into a covenant with himself (contrary to von Däniken's assertion that every religion had a 'covenant with God' — p. 73).

Enough has been said to vindicate those who have attempted to explain the wonders of Egypt as the flowering of a great human civilisation uninfluenced by any creatures from outer space. Many of the big questions presented by von Däniken as defying any answer acceptable to an 'intelligent' modern man are seen to lose their force once the full array of known historical facts have been set down and assessed objectively. It is refreshing to find so much interest aroused in examining the marvellous monuments preserved for us, as archaeologists, philologists, and historians keenly desire further light on much evidence meticulously assembled over the centuries. Of course it remains a possibility, but only a possibility, that earth has been visited by space travellers. In these days of secularisation, unknown or uncontrolled powers are not readily attributed to a special intervention of God. It is all the more imperative that the hypothesis of a helping hand from space travellers be subjected to the most searching scrutiny. Why should the known achievements of our race be ignored or belittled in order to transfer the credit to purely hypothetical astronauts?

BIBLIOGRAPHY

ALBRIGHT, W. F. *From Stone Age to Christianity*. N.Y., Anchor, 1957.

EDWARDS, I. E. S. *The Pyramids of Egypt*. Harmondsworth, Penguin, 1961.

EMERY, W. B. *Archaic Egypt*. Harmondsworth, Penguin, 1963.

GARDINER, A. *Egypt of the Pharoahs*. Lond., O.U.P., 1961.

FRANKFORT, H. *Kingship and the Gods*. Chic., Chic. U.P., 1962.

KRAMER, S. N. *History begins at Sumer*. N.Y., Anchor, 1959.

MONTET, P. *Eternal Egypt*. N.Y., Mentor, 1968.

WILSON, J. A. *The Culture of Ancient Egypt*. Chic., Phoenix Books, 1965.

A Funny Thing Happened on the Way to Mesopotamia

NOEL WEEKS

N. K. Weeks, M.A., Ph.D., B.Sc., Th.M., Lecturer in Ancient History, University of Sydney, has specialised in Mediterranean studies both in Australia and the United States. His work involved transaction of cuneiform texts, including the Epic of Gilgamesh, from which von Däniken quotes, and he is at present teaching Akkadian, the language of these texts.

The epic of Gilgamesh is generally regarded as one of the masterpieces of cuneiform literature. We are all aware that the masterpieces of our own literature are not as widely read as they might be and if that is the case then even less people will have read the Epic of Gilgamesh, even in translation. It is fortunate for von Däniken's thesis that he is able to cite such a little known source. Were the Epic well known then von Däniken's book would have had much less impact.

As von Däniken tells it[1] the third tablet of the epic relates how the the Sun God came with commotion and noise to seize Enkidu. He compares the weight of the Sun God upon Enkidu's body to the force experienced under rapid acceleration in a rocket. The fifth tablet is concerned with the visit of Gilgamesh and Enkidu to the tower in which the goddess Irninis lived. They were forbidden to look upon the face of the gods under pain of death. The seventh tablet tells of how Enkidu was carried aloft in the clutches of an eagle and how the earth appeared from the sky. In the eighth tablet we are told that Enkidu dies of a mysterious disease.

Von Däniken makes various comments on the significance of these details. He suggests that we have to deal with an exact account of the appearance of earth from outer space. There is one fact which undercuts the significance of all this speculation and casts an interesting light on the reliability of the author. These events which I have taken out of von Däniken's account of the Gilgamesh Epic *are not in the Gilgamesh Epic*. This can be quickly confirmed by consulting one of the numerous translations available.[2] Something vaguely like the supposed incident of the Sun God seizing Enkidu and lying upon him with great weight is found in Tablet seven 4:14-22. However the features crucial to von Däniken's argument such as the fact that it was the 'Sun God' and the 'Sun God's' great weight upon him are not present. Similarly, something similar to the story of Enkidu being carried up in the talons of an eagle is found in a completely different story, the story of Etana. Part of the relevant section of the story is missing and there may therefore be a question as to the exact translation. As von Däniken does not tell us by what warrant he restores and translates the text as he does one is at a loss to know where he derives the notion that the earth appeared like porridge and the sea like a water-trough. Certainly the text contains the idea that things appear different from the air but the text preserved hardly allows us to say that it appeared to Etana as it does to astronauts.[3]

This initial impression that the author has been careless or even irresponsible in his appeal to Mesopotamian evidence is confirmed by many other points. He claims[4] that the Sumerians had a highly advanced astronomy. As a matter of fact the Sumerians have left to us surprisingly little astronomical material compared to the later Babylonians. One list of about twenty-five stars has been recovered. This hardly amounts to great astronomical achievement. It is suggested that observations of the heavens were probably carried out by the Sumerians for purposes of fixing the calendar but none of these have been preserved.[5] As von Däniken gives no references I do not know where he obtained the information that they made very accurate estimations of the period of rotation of the moon.

One wonders at times whether the author has read some of the texts he cites. He says that cuneiform sources indicate where the remains of the boat used to escape the Flood should be.[6] He mentions as possible confirmation of this the fact that wood has been found at Mount Ararat. He then employs an extremely hypothetical argument to suggest that the Sumerian story of the Flood is genuine rather than the Biblical one. He seems unaware that the evidence he gives disproves his own argument. The Epic of Gilgamesh[7] gives the landing place of the Ark as Mount Nisir. This is a mountain somewhere in the Zagros range south of the Lower Zab river.[8] It is thus hundreds of miles south of the 'Mount Ararat' on which wood has been reported. If the remains of the Ark were to be found on this 'Mount Ararat' it would tend, if anything, to confirm the Biblical rather than the Mesopotamian account of the Flood.

The sweeping generalisations which the author makes about cuneiform literature finds no basis in fact. For example he says[9] that all the cuneiform texts from Ur tell about gods. These gods come from and return to the stars. They travel in ships in the heavens and have mighty weapons. No part of this description of the Ur texts finds any basis in fact. There were at least seventeen hundred tablets discovered at Ur which were not concerned with gods. There are texts concerned with business and commerce.[10] There are texts dealing with the sea trade.[11] Other texts deal with the thriving textile industry.[12] It need hardly be pointed out that texts dealing with the manufacture and sale of woollen garments are not concerned with gods and their frightening weapons.

Besides these business texts hundreds of literary and religious texts were discovered at Ur.[13] Even these do not conform to von Däniken's description. Many of the texts found are the same as those discovered in other excavations, e.g., Enki and Ninhursag, The Deluge, Gilgamesh and Agga, Gilgamesh and the Land of the Living, The Death of Gilgamesh, Inanna's Descent to the Nether World.[14]

A couple of tablets even describe life in a Sumerian school house. The subject matter of this text may be gathered from the first lines:

'Old school-grad where did you go (in your youth)?'

'I went to school.'

'What did you do in school?'

'I recited by tablet, ate my lunch . . .'[15]

This may strike some people as thrilling science fiction but it does not quite fit von Däniken's description of these texts.

Amongst other exciting literature I might mention the text which consists of instructions to a laundry man. To quote once again the first lines 'Come now, Cleaner, let me give you an order — clean my suit. The order which I give you don't lay aside.'[16]

I know of no cuneiform text from any city which would match the author's description of what occurs on every Ur text.[17] Cuneiform texts do not tell of gods coming from or returning to the stars. There is a very simple reason for this. The stars in Mesopotamian thought are not distant places but are manifestations of the gods themselves. They have personality and control various aspects of life on the earth.[18]

Since many of the gods who engage in battle are gods of wind and storm these are frequently the weapons they use. They also use conventional weapons like the net and arrow.[19]

The best way to determine how the Mesopotamians thought of their gods is to look at the pictures they made of them. The sun god indeed travels in a ship, but it is one that is propelled by punting poles and the artists have carefully depicted the water on which it is travelling.[20] Another cylinder seal shows a god travelling in a four wheel chariot.[21] The weapons used by the gods are the bow,[22] mace[23] and dagger.[24]

A star of some form frequently appears on these seals in association with the gods. This is usually in the form of an eight rayed star, but a four rayed star accompanied by the lunar crescent, or the lunar crescent alone is quite common.[25] Sometimes this is replaced by a central dot surrounded by a number of dots. These periferal dots are sometimes connected to the central one by lines.[26] The significance of this, as with many symbols, on these seals is not completely clear. However anyone who wants to explain them as planets circling a star would have to explain why only one orbit is shown. He must explain why as many as thirteen 'planets' occupy one orbit. In some cases the circle of dots even lacks the central 'star' and is carried in the hand.[27] Certainly these pictures show that there was a close association in the minds of the Mesopotamians between the gods and the stars. However, evidence that the gods were thought to come from the stars is lacking.

Examination of these same cylinder seals is the best answer for another erroneous claim by von Däniken. He says[28] that the Sumerian gods were not thought of in human form. On the contrary the many pictures and texts show us gods with very human shapes and attributes. As a matter of fact we are often at a loss to know whether a picture is meant to represent a man or a god. The best indication that a god was meant is often the presence of the divine crown. This crown is distinguished by nothing more technologically advanced than a pair of bull's horns.

Von Däniken does show a couple of cylinder seals.[30] He suggests that one of them shows a symbolic representation of an atom. I will leave it to the physicists to decide whether that is plausible. However a decision should take into account all the seals which show this or similar objects. Other representations[31] of this structure may not look much more like a tree but they certainly look less, if that is possible, like an atom.

The same seal shows something which von Däniken suggests could be an astronaut in a fiery chariot. I would suggest that the reader look at other seals with a similar figure[32] and make their own decision.

From the point of view of the Mesopotamian evidence this book is so full of error, mis-statement and untruth as to be worthless. I think that any reader who cares to read the texts and look at the pictures for himself will quickly come to the same opinion.

REFERENCES

1. *Chariots of the Gods?* pp. 65 ff.
2. Translations may be found in the following works: J. B. Pritchard (ed.) *Ancient Near Eastern Texts,* Princeton, Princeton University Press, 1955 or *The Ancient Near East,* Princeton, Princeton University Press, 1958; A. Heidel, *The Gilgamesh Epic and Old Testament Parallels,* Chicago, University of Chicago Press, 1946; N. K. Sanders, *The Epic of Gilgamesh,* Harmondsworth, Penguin, 1960.
3. For a translation of the Etana Myth see Pritchard, *Ancient Near Eastern Texts,* pp. 114-118. One should note also that Etana climbed on the eagle's back. He was not held in its talons.
4. p. 40.
5. KRAMER, S. N. *The Sumerians,* Chicago, University of Chicago Press, 1963. p. 90.
6. pp. 68 ff.
7. Tablet XI, line 140.
8. Heidel *Op. Cit.* p. 250.
9. p. 40.
10. GADD, C. J. *Babylonia c. 2120-1800 B.C.* Cambridge, Cambridge University Press, 1965. p. 8.

11. OPPENHEIM, A. L. 'The Seafaring Merchants of Ur', *Journal of the American Oriental Society*, 74 (1954), 6-17.
12. JACOBSEN, T. 'On the Textile Industry at Ur under Ibbi-Sin' in MORAN, W. L. (ed.), *Toward the Image of Tammuz and Other Essays on Mesopotamian History and Culture*, Cambridge, Harvard University Press, 1970. pp. 216-29.
13. For accounts of these see HALLO, W. W., 'Review of C. J. Gadd and S. N. Kramer, *Literary and Religious Texts*, Ur Excavations, Texts, Vol. VII', *Journal of Cuneiform Studies*, 20 (1966), 89-93 and S. N. Kramer, 'Literary Texts from Ur, Part II', *Iraq*, 25 (1963), 171-6.
14. Translations of these may be found in Pritchard, *Ancient Near Eastern Texts*, pp. 37-41, 42-57.
15. KRAMER, 'Literary Texts from Ur, VI, Part II', p. 174.
16. GADD, C. J., 'Two Sketches from the Life at Ur', *Iraq*, 25 (1963), 184.
17. The best descriptions of Sumerian literature and mythology, (besides the translations given in Pritchard, *Ancient Near Eastern Texts*), will be found in the works of S. N. Kramer, *The Sumerians* (see note 5 above); *History Begins at Sumer*, London, Thames and Hudson, 1958; *Sumerian Mythology*, Philadelphia, American Philosophical Society, 1944.
18. For an example see the ritual for the New Year Festival in Pritchard, *Ancient Near Eastern Texts*, p. 333.

    'The star Musirkeshda, who carries the (royal) scepter and circle, . . .
    The Eridu star, the possessor of wisdom, . . .
    Planet Jupiter, who carries the sign for all, . . .
    Planet Mercury, who causes it to rain, . . .
    Planet Saturn, star of justice and righteousness, . . .
    Planet Mars, fierce flame, . . .
    The star Sirius, who measures the waters of the sea, . . .'
19. *Ibid.* p. 67.

    'While the gods of battle sharpen their weapons.
    Then joined issue Tiamat and Marduk, wisest of gods.
    They strove in single combat, locked in battle.
    The lord spread out his net to enfold her,
    The Evil Wind, which followed behind, he let loose in her face.
    When Tiamat opened her mouth to consume him,
    He drove in the Evil Wind that she close not her lips.
    As the fierce winds charged her belly.
    Her body was distended and her mouth was wide open.
    He released the arrow, it tore her belly,
    It cut through her insides, splitting the heart.'
20. See Pritchard, *The Ancient Near East*, picture 169. H. Frankfort, *Cylinder Seals*, London, Macmillan & Co., 1939, plate XIX, XV.

21. Frankfort, plate XXII.
22. *Ibid.*, plates XIX, XXII.
23. *Ibid.*, plate XIX.
24. *Ibid.*, plate XVII.
25. *Ibid.*, plates XX-XXXI.
26. *Ibid.*, plates XXX, XXXI. See for discussion of these dot patterns E. D. Van Buren, "The Seven Dots in Mesopotamian Art and their Meaning", *Archiv fuer Orientforschung*, 13(1939-41), 277-89.
27. Frankfort, plate XXXIII.
28. p. 42.
29. Frankfort, p. 32.
30. His plates are not numbered but this would be on plate 12.
31. Frankfort, plates XXXII, XXXIII,
XXXV, XXXVI.
32. *Ibid.*, plate XXXIII.

An Engineer's Opinion

CLIVE HOULSBY
Chartered Civil Engineer (Australia)

A. C. Houlsby is a practical engineer. He spends his time travelling throughout N.S.W. checking and building dams for the Government. He is a 'grouter' who pumps cement under great pressure into defective structures.

'The crew, Mycerinus is drunk'
'The crew, Cheops excites love'
'This side up'
'To be taken away'
'For the royal tomb'

These are graffiti from quarries from which stone for Cheops Pyramid was taken.[1] They do not seem to suggest that Egyptian builders were significantly non-human, as Erik von Däniken seems to believe.

The first two inscriptions are the cheer-leaders' chants for two of the crews working on the job (each crew was about 1,000 men). Somehow I fail to be inspired by their literal meanings—but then maybe the ancient pyramidians (to coin a word) knew some naughty stories belonging to each chant.

The other three inscriptions are destination signs on blocks. All the inscriptions are signs of a very ordinary human kind of labour. Von Däniken has the impression that these blocks were about 12 tons each and he is perturbed at the difficulty of transporting them (distances were up to 600 miles). Perhaps he forgot to do his field-work, because we have been assured by those voyagers who have actually hazarded a trip to the Near East, complete with tape measure, that nearly all the blocks are only 2½ tons.[2] A few special ones are much heavier than this.

Now 2½-ton blocks are not so very difficult to move given determination and grit. With the aid of carefully graded tracks nicely lubricated with grease from the mess kitchen and gangs of healthy coerced volunteers, it wouldn't take anywhere near the several hundred thousand fellows presumed by our friend (p. 101) nor even the 100,000 of Herodotus' tale.

For the probable conditions, the coefficient of static friction would range from 0.30 to 0.65[3] requiring about 20 men to manipulate each block, under average conditions. It seems[4] that there was a permanent staff of 4,000 specialists (their barracks have been found!) augmented by gentlemen rendered unemployed by the seasonal floodings of the Nile; possibly there were less than 50,000 of them.

Quarrying required patience and perseverence. However, the copper and timber wedges mentioned by von Däniken (p. 85) were only part of the tools available at

Stepped Pyramid of King Zoser. Sakkara.

that time.[2,5] Moreover the step pyramid of Sakkara is mainly limestone and not granite as von Däniken thinks. Limestone is easier to dress and handle than granite.

A few of the many other clangers with which our author's works resound include—

(a) no modern architect (he means engineer, of course!) could build a copy of the Cheops Pyramid (p. 100). We could, if we were allowed to get stuck into it.

(b) 664 years would be required to build a pyramid (p. 101). It could easily have been done within the lifetime of a Pharoah.

(c) ropes were non-existent (p. 101). They were not.[2]

(d) his allusions to the supposedly 'classic' proportions relating to the distance to the sun, the balance between continents and oceans, and the mathematical proportions. These concocted phenomena originated with Taylor and Smyth, last century, and have long been discredited.[6] Their proponents have been dubbed 'Pyramidiots' (p. 179).

(e) his diatribe at p. 97 about the use of rollers is quite unnecessary (skids were á la mode in those days). His offer of his kingdom for a convincing explanation of the transport of the stone blocks is accepted with gusto!

Nor must we gloss over a major soil mechanics blunder concerning an imagined airfield at Nazca, in Peru (p. 32 and several pictures). The plain here has a number of depressions incised in it to a depth of about one inch. The depressions, being generally straight, and probably artificial, have been postulated as airfield markings. However, the surface is not rock! It is a soil[7] with apparently the typical loose dispersive structure of arid zone soils; at the first blast of slipstream from plane, flying saucer, or what ever, the surface would have blown away smartly. The surface is so soft that a single passage of a cart or car across it, apparently leaves tracks discernible for many years after (perhaps centuries!)

The ancient engineers built many superb structures in addition to those selected by our imperturbable friend (his selection seems biased towards the mysterious). Some of these other works included enormous dams, canals, walls, wells and so forth,[8,9] the very mundaneness of which tend to negate his theory of 'astronaut' intervention.

It seems fatuous of Herr von Däniken to have built such an extensive theory on such readily disprovable premises. He should have done his homework better!

REFERENCES

1. COTTRELL, L. *The Mountains of Pharoah.* Lond., Pan. p. 252.
2. DE CAMP, L. S. *The Ancient Engineers.* Adel., Rigby. pp. 40, 42.
3. ESBACH, O. W. *Handbook of Engineering Fundamentals.* N.Y., Wiley. pp. 4-52.
4. DE CAMP, L. S. *The Ancient Engineers.* Adel., Rigby. p. 41.
5. COTTRELL, L. *The Mountains of Pharoah.* Lond., Pan. p. 85.
6. IBID. p. 178 ff.
7. HOWELL, M. and MORRISON, T. *Steps to a Fortune.* Lond., Bles. p. 15.
8. BISWAS, A. K. *Hydrologic Engineering prior to 600 B.C.* in *Journal of Hydraulics Division,* volumes HY5 1967, HY2, 1968, HY3 1968, HY5 1968, HY2 1969. American Society of Civil Engineers.
9. MANER, A. W. *Public Works in Ancient Mesopotamia* in *Civil Engineering,* July, 1966. p. 50 American Society of Civil Engineers.

Riders to the Chariot

JOHN GENT
Chaplain, Pulteney Grammar School, Adelaide, S.A.

The Reverend J. G. M. Gent, B.A. (Adel.), Th.Schol. (A.C.T.) is a well-known contributor to theological conferences and journals. An Anglican, he has been a parish priest, and has had very extensive experience as a school-master.

Although its viewpoint is Christian, this paper is not intended, except perhaps indirectly, to be a theological or historical criticism of *Chariots of the Gods?* Its object is more difficult and more ambitious; it asks 'What is the reason for the extraordinary success of this book?' Or, more largely 'What is the reason for the successes of books *of this sort?*' for *Chariots of the Gods?* is one of a kind with, for example, *Collision between Worlds* and *The Sacred Mushroom.*

This is an important question to ask. Here is a book which does not, and could not, stand up to an hour's examination by an historian, a philosopher or a theologian; and yet it has become a race-away best-seller, exerting phenomenal influence over a considerable number of people as history, philosophy and as theology.

It is not possible here to give anything but hesitant, subjective answers; the present writer has neither the opportunity nor the expertness to undertake a definite piece of sociological research. It is important that the question should be posed and an answer at least attempted.

A century of free, compulsory, secular education (which naughtily, Arnold Tonybee called 'gratuitous, enforced, godless instruction') has produced in the West a society which is technically literate. But literacy is not necessarily education, and it could be alleged that there is a sad lack of depth and of discipline in contemporary education.

The great emphasis in the education required for a technocratic society has been upon the natural sciences, and especially upon the applied sciences. Those of us who have had much to do with scientists do not need C. P. Snow to point out a certain naivety in human understanding judgments, and in broad cultural judgments, which seems not infrequently to accompany proficiency in the natural sciences. But, such is the mystique which attaches to 'science' and its 'certainties', and such too is the undeniable pragmatic success of the 'scientific method' that three consequences of present importance can be suggested.

i. Anything not 'scientific' can be dismissed as subjective and incapable of proof. This applies to history as a study, to literature, to the arts, and most certainly to theology.

ii. Any work which can make some claim to be 'scientific' or to exhibit, even speciously, 'the scientific method' can hope to be accepted by the literate non-scientist as objective and certain knowledge. (Notice, for example, the overt appeal of what is called 'Scientology'.)

iii. There is amongst the literate public a gross ignorance, not only of the 'facts' of history, but also of the methods, techniques and possibilities of historical study. Indeed, there may be said to be a widespread ignorance of the ways in which the humanities are to be studied and how they have organised themselves.

A book such as *Chariots of the Gods?*, which the unconverted might describe as science-fiction dressed up as history, is sadly likely to appeal to readers who have been allowed to develop neither the critical skills nor the knowledge to scrutinise it with appropriate objectivity. This was true of the not dissimilar *The Sacred Mushroom,* whose refutation (as by Robert Graves) demanded knowledge within a relatively obscure and specialised field.

But there is more to it than imperfect education. Most of us would recognise in ourselves something of an attraction towards two elements of human experience: the esoteric and the occult. By 'the esoteric' I mean the knowledge which establishes one as a member of an inner group, as one of the people who 'really count' and have powers which the common people of the outer group do not share. (C. S. Lewis had some wise and chastening remarks on this in his *That Hideous Strength.*)

The 'occult' is similar and different. It is knowledge which lies hidden in shadowy realms behind locked doors, known only to a few, who may, however, permit us to join them and to have the key. The continuing appearance of advertisements for organisations like the Rosicrucians is evidence that the pull towards the occult can be commercially tapped. At its most authentic, occultism is the sickness of Faust, the search for one secret that solves everything.

Is it fair to say of the avid readers of *Chariots of the Gods?* not only that they are imperfectly educated but also that they submit too readily to the attractions of the esoteric and the occult? Perhaps. There is, after all, much in the book that is plausible, and, granted Erik von Däniken's assumptions, there is a moderately coherent scheme. But what strikes the present writer most forcibly in his own reading of the book is the appeal it is patently designed to have in the esoteric-occult area. To climbers on to the Chariot-band-wagon it offers these three clear satisfactions:

i. They have *the* key.
ii. They are illuminati, possessing knowledge beyond that of the common man.
iii. They are constantly and explicitly assured that those who disagree are reactionary, close-minded and arrogant.

But even this may not be all.

For several centuries in the West there has been an increasing resistance to man's sense of the supernatural. There have been attempts, many of them brilliant intellectual constructions, to explain away man's religious impulse. It has been dismissed as a relic of man's earlier ignorance, to be evaporated by the rising sun of Reason. It has been explained as something encouraged and inculcated by power-elites to be 'the opiate of the masses'. It has been explained, for example by Freud, as the consequence of unfunctional family relationships. Latterly, it has been described in terms of Existentialist *Angst*.

And yet, every time religiousness has been explained or exploded, there has come into being new 'surrogates for religion' as Aldous Huxley called them: the State, the party, the super-ego, commitment, involvement . . .

There have been two aspects of our resistance to the religious impulse, inextricably intertwined in reality but separable in theory.

First, there is a cultural aversion to religion. It expresses itself as an opposition to authoritarianism or 'the structured church'. It rejects hierarchies and (on scientific grounds) the images and apparatus of 'organised religion', especially those pastoral images from the Bible which seem to agree least with urbanized technological society. Curiously, this thinking resents the priest in his vestments of office but venerates the scientist in *his* vestment — the white coat, necessary even in tooth-paste advertisements, which implies the authority of a structured body called 'Science'.

Secondly, aversion to religion may be personal, for men wish to establish human autonomy, whose exercise makes less pressing demands than 'religion'.

It is here that one nearly comes to despair. In von Däniken's *Chariots* the whole traditional notion of religion is queried, yet there comes in its stead something which is at least quasi-numinous. The earth, it is suggested, has been visited by beings which are 'other', superior in wisdom, intelligence and technique. The thought of them makes the flesh creep, pleasantly and romantically.

The 'numinous', the 'awe-ful', is what our concrete ant-heaps would destroy. Rationalism and materialism have denied man his traditional expressions of wonder and awe. And yet the UFOs keep coming, strange and unpredictable. And not only do they come now, they are supposed to have come in prehistoric times.

Whether such objects exist I do not know; but I think I know enough of modern man and his needs to say that if they did not exist he would have had to invent them. In the news of space-visitors we have the new images, the new symbols, of man's deep sense that he has always been subject to visitation. The Christian would say that indeed he has been, and is, subject to visitation. But not by trans-stellar invaders.

Every myth is somehow an anticipation or a reflection of the reality. There are ways in which *Chariots of the Gods?* points to the reality of the Incarnation, a much more remarkable event than any of those which it imagines, and one which is very precisely and circumstantially located in history. There *were,* we are told, space travellers around Bethlehem at the time, and they are reported to have spoken to some shepherds. But their function on that occasion was secondary.

G. K. Chesterton said, many years ago, that when men give up belief in God they do not believe in *nothing:* they believe *anything.* The fascinating and annoying thing is that human myths always suggest, somehow, the truth. And that may be why a book like this has its appeal. Its enthusiasts may intuitively recognise that it is all a bit, a little bit, like what must have happened.

The more prosaic the world becomes the more men look for poetry. They do not call what they look for poetry, for 'poetry' is not 'true', truth being what can be weighed, counted and measured. When the old myths are said to have been discredited, man makes new ones. Von Däniken can see that primitive and ancient man was a myth-maker: but absolutely and humourlessly he fails to see that he himself is trying to devise a new myth, a kind of scientific cargo-cult. Speaking of the 'boundless stupidity' of which people are capable, he says: 'There are groups who develop fanciful religious ideas from hitherto unexplained phenomena or build cranky philosophies of life from them.'

He would deny that he himself was developing 'fanciful religious ideas' from inexplicable phenomena. And yet everyone of his readers is being invited by rhetorical question after rhetorical question to join such a group and to make just that sort of development.

Of course the book is popular. It speaks of the numinous, of the unearthly, and of mysterious visitors. It encourages its readers to make a new version of the persistent myth which points, like a rumour, to the real and eternal Visitor.

Insofar as *Chariots of the Gods?* states, or suggests, doubts as to the validity of the main items of Christian doctrine, rebuttal is easy. Christian people will not be troubled by it. The Christian faith is anchored firmly in real history. The New Testament is not set in a misty background of primitive society whose mental categories are inaccessible to us; it belongs to a historical milieu which can be studied and known in the most intimate detail by the clear light of ordered historical research. And space travel does not come into it at all.

How to be a Leader of Opinion

H. D. NICOLSON
Ancient History Master, Cranbrook School.

H. D. Nicolson, B.A., is a Master at Cranbrook School and a Part-time tutor in the Dept. of Adult Education, University of Sydney. He is a well-known author of critical and historical works, lecturer and controversialist.

With about the same regularity that cargo cults spring up in developing countries today, religious revelations concerning the imminent destruction of the world or concerning extraordinary advents used to be made in countries affected by European civilization in the last century.

Why didn't the leaders of the society put down the leaders of these movements? Apparently, no amount of learned criticism or official disapproval prevented these 'revelations' from becoming popular. Even after the predictions of the 'stirrers' were proved false people did not lose faith altogether. As there were references in sacred texts to cataclysmic happenings, then cataclysmic happenings were to be expected. The notions were true: human prophets might be mistaken as to timing. The heart continued to long for a new world even when the eye could not see any signs of change.

Then, as always, the general body of theory and philosophy seemed not to reflect the interest that the public at large took in such revelations. Theorists as individuals do feel sympathy with the hopes and fears of their fellow humans. Quite often, indeed, they feel a deep compassion for those suffering from a sense of frustration and a longing that SOMETHING should happen.

Because most educated people used to be steeped in classical history they easily recognised, even disguised in modern language and modern forms, the portents, cults, images and formulas which had sprung up in all ages to satisfy the perennial human hunger for change, for the marvellous and for the overwhelmingly convincing. But to most philosophers the persistence of the hunger seemed a more important matter than any of its obviously limited forms of expression.

'What,' they asked themselves. 'was the ultimate goal of the striving of mankind?' And they tried to see, 'as through a glass darkly', the object or objects of the hunger.

In the nineteenth century a philosophic movement called Absolute-Idealism was most thorough in working out the implications for theory and for life of this and related questions. It stands to the credit of the Absolute-Idealists that they revealed the difficulties involved in such an enterprise. But even if there had not been irresolvable intellectual problems, the ordinary man did not recognise that philosophers were trying to meet *his* needs, and he found the Idealists too remote.

The one group of philosophers who managed to speak to the literate 'ordinary' man were the pragmatists, of whom the chief was probably William James.

Many of the phrases, if not the actual arguments, of William James's essays passed into general use. It was widely assumed that he, particularly through his *The Will to Believe,* had provided a solid intellectual basis for the view that the social and political world can be altered by belief, and that this was true regardless of the objective worth of the belief in question.

The older religious view that faith could move mountains was based on faith in a universal power. James's observations suggested that it was the faith *itself* that did the work. Not only as a result of James's studies, but also because of a weakening of faith in the general order, this century has seen many individuals and groups trying to prove their own will against all comers. Moreover, when such individuals or groups fail (as the Nazis failed), they often attribute their failures not to the limiting conditions of their type of activity but to their own failure to be daring enough, or ruthless enough, in the execution of the programme. In brief, they say they fail because they have not *believed* strongly enough.

Now the use of the idea of the daring as though it were a mark of all intellectual progress, and not just an accidental relationship, in which some intellectual enterprises become involved, has been a mark of sophistry through the ages.

The name 'sophist' was first applied by Plato who established in Athens in the fourth century B.C. the first regular centre for the pursuit of enquiry. He and his followers first noted and defined that class of men who wanted to learn so that they and their outlook could have dominance. The sophists taught the young how to win arguments, votes and popular appeal. As Plato put it, they were not so much concerned with truth as with the image or appearance of truth.

From Plato's day to this one of the philosopher's secondary tasks has been to distinguish between philosophy and science on the one hand and sophistry, or truth-as-it-serves-the-purpose on the other. It follows from the character of sophistry that its outward character will change from age to age, that what can be contrived to impress an age brought up on only a few years of simple schooling will not impress an age which is rehearsed in the physical sciences and kept up to date by television. But it remains true that philosophy and learning have not only to pursue their own learnings and teaching programmes but also to defend themselves from spurious studies and spurious sciences.

We should not expect that the hunger for the marvellous or the overwhelmingly convincing will be wholly assuaged by school-teaching and television. To my knowledge there has been no successful widespread religious movement proclaiming a revelation of changes in the state of the world since World War II. The astute propagandist for any creed today will be careful not to offend modern susceptibilities by referring to things which can be easily checked by people with reference libraries or university-trained authorities close at hand: he will try to refer to generalized wishes and future plans that people seem to accept as belonging to them as persons but which are outside the range of critical analysis, such as the desire for personal experience of excitement, travel, food, sex and so on.

The role of propagandist is a secondary one for the man who desires to be recognised as a Leader of Opinion. To carry authority such was enjoyed by the sophist in classical societies, the witch-doctor in simple societies, the prophet in religious groups, the pace-setter in the twentieth century technological society has set himself a difficult task. He does not have to be ingratiating like the propagandist, but unsettling; he does not avoid physical science, but creates the impression of knowing all about it and going beyond it; he does not have to consider difficulties, but in drawing up his plans for the future casts his net ever wider, so that any difficulty that pops up can be accepted yet dismissed as too trivial to disturb a brilliant idea. In a nutshell, the modern sophist must be *daring*.

There are, however, limits to daring. The modern Leader of Opinion is leading men. He must not therefore be mistaken for the leader of ideas.

If the leader is just a scientist, he can go about his business by doing science. It is important, therefore, that the Leader of Opinion should have insight first. If he has that it is better that he should be self-taught. His insight will tell him what science to learn and what to reject as well as telling him how to lead scientists.

If the leader is just religious, he will already have the greater part of the faith behind him, and that will not leave him the essential, originating, leadership role. If a leader is to lead it is better that everyone trusts to his inspiration. After all, the great discoveries of the past were made by men who challenged accepted opinion. The leader's opinion can always be assured of being the most unusual one by simple device of incorporating into it any new and current speculation, or by associating with its author and speaking of him as though he were a colleague.

As to philosophy and philosophers, it is safe to assume that they do nothing. What has the public to gain by referring to arguments over logic, linguistic classifications, inter-disciplinary studies, theory of science, or generalizations about categories? The public wants reassurance about its prospects for the future. It wants experiment on a scale suited to its needs and designed for the largest possible range. Old truths are inadequate. Argument divides us and wastes time and words. Let us persuade men to embark on a positive exercise.

As the Leader of Opinion must not be taken for a conventional leader of ideas, so he must not be mistaken for the conventional leaders of men, politicians and business men.

If he were just a politician he might be responsible for details which would detract from the whole. Moreover he would be limited to one country. The ideal Leader of Opinion will go to places where he can find the secrets of the future and to the people who have helped to unwrap some of the mysteries.

To be like a business man, concerned with profits and not willing to risk time and interest on the unusual and the long range, is obviously a mistake. However business, with its emphasis on enterprise, publicity and being photographed and associated with long-established and well known wonders of the world, recorded in hundreds of encyclopaedias and children's educational books, can obviously teach something.

Thus by careful selection and daring espousal, by leaving out all that might hinder or detain, the Leader of Opinion can summon up a vast picture of the future and show all the marvellous portents in the past. Does it matter that we have to imply that scientists and philosophers are not getting on with the job? Does it matter that thousands, who have not the resources to follow up the many loose ends, are bamboozled? If the leader is listened to, people will learn something, if not, the leader is able to parade as a neglected seer and prophet. The leader can always enjoy the ascription to himself of the title of courageous.

Why should we look out for such a sophist? Probably because in the end civilization can only bear so much sense of insecurity about the sources of truth. If scepticism about learning and truth becomes too widespread, people may withdraw from commitment.

A LETTER

My dear Editors,

Thank you for your kind words about the essay I wrote after you sent me Von Däniken for review. I had thought when you asked me to comment on it from a philosophical point of view that it would raise certain issues worthy of comment; but I had to write in the end an essay warning against the spread of this sort of sophistry because my view is that notoriety in this age means publicity: and publicity, good or bad, means financial success and further productions of the same kind. I have even seen advertisements, which say, "Come and hear the controversial, Mr. X! Judge for yourselves!" — not about this book but similar sorts of things.

It should be made quite clear that the book raises no philosophical issues whatever. It does raise some questions of method.

The main suggestion of the book, that some civilizations were given an extraordinary impetus by the arrival of spacemen from other planets, raises more problems than it solves. However, these problems are scientific problems not philosophic, because they raise no fundamental questions such as those about the nature of facts, only further questions about where and what to look for when accounting for the rise of civilizations. If a civilization is given an impetus by inter-planetary visitors, they must have come from other civilizations. Did these civilizations have to be given an initial impetus also? Did no advanced civilization arise from causes that were to be found on its home planet? If not, do we have to look for an indefinite number of planets with civilizations each started off by civilizations elsewhere?

There are other general tests that we can apply to the hypothesis that civilizations on this planet were affected by visiting astronauts. First, which civilizations should be affected most: the arrested, the terminated, or the advancing technological ones? It is strange that most of the evidence for extra-terrestrial visitors has been deduced from arrested or terminated civilizations. Modern western European civilizations do not seem to have had the honour of visits for a long time. In their case it is sometimes explained that modern scientific thinking and technological practices have been, unknown to those concerned, affected by extra-sensory communicating influences. This last additional hypothesis may meet with opposition from some physical scientists, who will not like being compared to spiritual mediums. Some scientists may not mind. Sir Arthur Eddington and Sir James Jeans earlier in this century supported spiritualist interpretations but met criticism from many others including Professor L. Susan Stebbing in her well-known work *Philosophy and the Physicist*.

Whether the impetus to civilization is conveyed directly by visits or by some kind of extra-sensory communication, the problem still remains the same. You have to account for two civilizations now instead of one, and you still have to find out how the first began. Did it begin by natural causes, covered by theories such as evolution, trial-and-error, diffusion of culture and all that, or did it come from another planet and so on ad infinitum?

And the problem is worse than that because you have to give an account of the encounter between the two civilizations. We have one set of visitors (who are largely unsuccessful at communicating) and another set of communications which are masquerading as discoveries and inventions made by brilliant men, but are really extra-sensory suggestions falling on prepared ground. And there is no possible way of saying when one or the other of these variables comes into operation. If a test is to be found it must be a scientific empirical (i.e., observational) one based on occurrences in some stated time relationship.

To conclude this section: when philosophers are trying to give an account of the logical properties of argument, or scientists are trying to give an explanation of occurrences, they tend to favour the explanation which uses the minimum number of terms or entities. 'Interventionist' theorists have to justify their views with evidence if they wish to explain changes in world history by intervention of new entities from outside the world. I shall turn now to consider the feasibility of the evidence of the theorists who favour interventionist views. The evidence would have to stand independently of any reference to the entities whose interference it seeks to prove.

If I may intrude on the ground of the cultural anthropologists, archaeologists and prehistorians in order to illustrate my methodical point about the economy of the hypotheses or the minimum number of explanatory entities, let us look at the types of explanation of such things as great stone (megalithic) monuments, symbolic designs on ancient monuments or the handing down of laws, myths and legends. There are easily recognisable five main types of explanation:

1. natural interraction of humans, animals, plants and physical things
2. intervention from above the earth
3. intervention from below the earth
4. extra-dimensional intervention (e.g. spiritual, E.S.P.)
5. combination of the above.

Looking at explanation of the unusual feature of megalithic monuments such as the pyramids of Egypt, the altar of heaven at Peking, the Borobudur of Java and the step pyramids of Central America the first or natural explanation tends to be in terms of such factors as surplus food supplies (generally supplied by advance to agriculture and herding), centralized governments, population increase, periodicity of work and ritual reinforcement of social demands.

All the other explanations have the additional tasks of accounting for the interaction of the more advanced intervening force with the less advanced; for the reason for the withdrawal of the intervening force; for the lack of economy of the expression shown by the intervening force; for the variety of styles manifested in the megalithic monuments themselves, which calls for separate as well as unified treatment; and for other problems. If the intervening force only accounts for some of the monuments or only for the marvellous features that appear in some of them, then we shall still need "1. natural interaction" as the explanation most of the time, and we shall only call in the other form/s of intervention when we cannot see how to give an account in terms of the first.

As to the question of designs found in connection with ancient monuments indicating 2, 3 or 4 I feel that the evidence is split. For every design showing some intervention from above the earth I can think of evidence for intervention from below. For every representation that can possibly be read as a spaceman I can think of another that represents man as a serpent-like creature or a fish. As to design trying to represent the shadow, the spirit or the breath of man they are numerous and colourful. Any book which concentrates on one or other of these categories could easily mislead those who have not seen directly or by illustration the monuments and designs left by ancient man. Any explanation of ancient man's activities would have to fit in a general framework of explorations which would explain *all*. Australians ought not to forget the pioneering and still useful work and exploration in this field of their fellow country-man, the late Gordon Childe, whose two books *What Happened in History* (Penguin) and *Social Evolution* (Fontana) are readily available.

To turn now to a completely different methodological issue. If there was technological intervention in man's prehistory by inter-planetary travellers the only way of judging that intervention is from the view-point of technology as it is at present developed.

If objects were found which modern technology could not duplicate that does not necessarily mean that those objects have the properties that they do have because of some other technology about which we do not know. In reverse, if miracle blue prints were suddenly to descend on us, until our technologists could handle them they might even divert the more steady and reliable development of technology. That is to say that it is only by understanding the technology we have that we can hope to develop, and the questions of how the present set of practices and theory are to advance are open ones determined by a large number of factors, so that even the possibility of discovering extra-terrestrial technology would have to take its place as one among a number of objectives. There is some evidence to suggest that the interaction of two technologically oriented cultures does not necessarily stimulate the weaker of the two and that it could quite easily inhibit it.

May I conclude by saying that no book arguing for intervention from beings above or below the earth or extra-dimensionally has ever helped me much to appreciate the quality and causes of the matters it sought to illuminate. The particular book you sent me with its accent on marvels rather than the steady evolution of man in small matters such as spinning thread, making bread, domesticating animals, enjoying beautiful or pretty things such as trinkets for children and jewels for women, I found too reminiscent of my childhood encyclopaedias and Wonder Books of Long Ago to seem real. Big things are quite often mistakes and the Marvels of the Fair are often not so marvellous when you see them from behind.

Your truly,
H. D. NICOLSON.

Chariots on the Wide Screen

EDGAR CASTLE
Senior English Master, Cranbrook School, Sydney

Chariots of the Gods? has been popular for reasons which are not hard to find. Among them will be that it provides the chance of a holiday from the conventional wisdom of our time. Like the 18th century Gothic novel, it offers to those willing to suspend their disbelief thrills which rationality cannot.

Like the Gothic novel, however, it also has a proportion of its readers (in this case possibly a large proportion) whose disbelief is not easily re-engaged. It is fear for them, and conceivably *of* them, that has prompted this present volume. Nevertheless, it is to be believed that truth *is* great and *must* prevail. And it is to be expected that Erich von Däniken's hypothesis will one day join Joanna Southcott's box in the limbo of forgotten messianisms.

In the case of the many thousands of readers of the book, the victory of truth over nonsense can be expected to be complete in about a decade. In the case of the millions who have seen the film made from it, it is not so easy to be optimistic. It is about the film that this essay is written.

Chariots of the Gods (the film title is subtly different from the book title) was produced in 1969 by Terra Filmkust of Berlin. Its director was Dr. Harald Reinl, a talented director in the tradition of Leni Riefenstahl. Most of its footage is expensive on-location work. It had a specially composed musical score played by a full orchestra. (The film music is now commercially available on discs.)

Most of the material filmed is strictly 'documentary', though the opening sequence — of a cargo-cult ceremony —is so smoothly done that it seems to have been set up. Overall, *Chariots of the Gods* can be described as an expensive, elegant and effective piece of conservative cinema.

Dr. Reinl has been quite explicit about his aims. In a press release issued by the film's Australian distributor, he is reported to have said, 'The film is nothing but an hypothesis. It does not pretend that is how it was, but says that is how it could have been.'

I must say at once that I do not know how an hypothesis

can be presented in any film more ambitious than an educational animation. For the business of the film is illusion and its aim is the total involvement of the audience with illusion. I can no more imagine an hypothetical film that I can imagine a love poem that 'does not pretend that is how it was'.

The Japanese film *Rashomon,* some years ago, set out to tell a story in four versions, each of them 'how it could have been'. That, possibly, could be called a group of hypotheses. But in this, as in any film of quality, what 'could have been' took on the lineaments of truth and became 'how it was'. The whole point of *Rashomon* was that the audience should be driven to distraction by trying to find the one 'true' version of four equally artistic film tales.

In the case of *Chariots of the Gods* Dr. Reinl cannot have set out to do anything less than a director's best; and that would, in this case, have been to point up the mysteriousness of von Däniken's 'mysteries' and to pose, with cinematic force, von Däniken's astounding rhetorical questions. In a phrase, to dramatise von Däniken. Hypotheses are usually modestly stated, and do not need widescreen colour and special music.

Undoubtedly, Dr. Reinl and his team worked with artistry and integrity. Their film, in its relation to von Däniken's text, raises a number of questions about 'documentary film' in general.

One, for example, is a new and urgent version of the old journalistic dilemma: the truth, or a 'good story'?

Another is whether *any* kind of tentativeness is transmissible by film or television. Von Däniken's text has the appearance of speculation. The film cannot by its very nature be speculative. What it shows must seem to be true, at least at the time. If McLuhan is right about the importance of the electronic media, then it will soon be the case that what the community believes about anything will be — like a sailor's love-affairs — temporary, but never, never, tentative.

However all this may be, the film *Chariots of the Gods* is a beautifully made soft-sell for the *Chariots of the Gods?* thesis and it is dangerous.

The book is not dangerous, because it is in an intellectual mode to which we are well accustomed. It is sequential. It has propositions in print. It has the appearance of logical structures whose rules we know. We are able to say of the book that it is mistaken in such a sentence on such a page, or that it draws an invalid conclusion in another place. Indeed, you can do to the book just what most of my co-contributors have already done.

The film, however, is a subtle and complex artistic entity that will not sit still to be examined. It leaves a total impression, a large part of which is subliminal. It does not allow its factual and rational components to be separated from its rhetorical devices. As it is a technically good film, it leaves an impression that will be remarkably persistent. And since individual details fade from the mind, and names are rapidly forgotten, it is proof against close logical analysis.

The popular success of *Chariots of the Gods* is, in its way, sinister. Hundreds of thousands of people have been to some degree persuaded by it that space-travellers originated human culture, if not humanity itself. Of course they were already open to that conviction. But, without the film, they might also have been much more accessible to the contrary view.

There is a Kipling story about a village that voted that the earth was flat. I am inclined to fear that it is possible that, because some film-makers and film-distributors see a possibility for profit, we shall see the day when the Global Village votes for something equally silly.

I do not begin to know what can be done. Censorship is unthinkable. The rule *caveat emptor* must run for films, even though it is expected to run less and less for consumer goods. School lessons in film criticism are able to teach young people to recognise the cruder devices of film rhetoric: but *Chariots of the Gods* was not crude.

Perhaps the only thing to hope is that the community will develop the only skill, according to T. S. Eliot, likely to be effective in artistic criticism, which is 'to be as intelligent as possible'.

'Excuse me Ezekiel'

ALAN COLE

The Rev. Dr. R. A. Cole, M.Th. (Lond.), Ph.D. (Dublin), is the Head of Department of Old Testament Languages and Literature, Board of Divinity Studies, University of Sydney.

Does the prophet Ezekiel really describe the landing of a spaceship (or a wheeled multipurpose space craft) in Chapter one of the book in the Old Testament that bears his name? So von Däniken suggests on pp. 55-7 of *Chariots of the Gods?;* and certainly we do have wind, noise, fire, wheels, and metallic gleam described. It would indeed be fascinating if we had here a contemporary account of the landing of some actual spacecraft: but Ezekiel himself describes what he saw as 'visions' (Ezek. 1:1) rather than objects. Indeed, he uses stereotyped visionary phrases like 'the heavens were opened' (Ezek. 1:1) and 'the hand of the Lord was upon him' to make absolutely plain that what follows is to be seen in the sense of an ecstatic mental experience, rather than sober scientific description of an event. It is true that there is here a description of some kind of sacred chariot, such as Ezekiel may well have seen used in the processions of the idols of Babylon: but it is also true that the description is not self-consistent. All is well: this is the stuff that dreams and poetry are made of: we do not look for logical consistency in such areas. But the few details (like fire, noise, light, and wheels), that might fit a hypothetical space-craft, are not the whole of the description: it culminates, not in wheels or in chariot, but in a great throne set above the chariot (Ezek. 1:26), and God, in human form enthroned there. If we take the chariot literally, then all of this, too, must be taken literally.

Interpretation

And what of the four 'living creatures' that draw the divine chariot? Composite and artificial beasts like these (winged, with faces of man, lion, ox, and eagle) can be paralleled from Western Asia archaeology as statues of 'temple guardians', symbolising the ministrant spirits that attended the gods, in men's eyes. The Sphinx in Egypt was such a composite figure: the Cherubim that shaded the ark underneath their wings (Ex. 25:20) were probably also of this composite form, so that such beasts were not strange to Ezekiel. But what place have they in a scientific account of a space-ship? And, if it be objected that these are accretions and irrelevant, then why is the description of the divine chariot itself (the so-called space vehicle) not equally an accretion and irrelevant? Both parts of the description stand or fall together. On the other hand, if these are indeed visionary experiences of some kind, as the author himself claims, what more natural than they should be a very medley of bizarre shapes either seen by him in statuary form in the waking world, or familar to him from tradition?

The Psychology of Ezekiel

Psychologically speaking, we always build the unknown from the known, and this would be another example of the same principle. But perhaps our author was not the man to have visionary experiences? Indeed he was: so prone is he to visions that men have wondered if he were not an epileptic. While we may not adhere to this view, it is obvious that the psychology of Ezekiel was very unusual. Besides a repetition of the chariot vision of chapter 1 in chapter 10 (where it is now sited in Jerusalem, not in Babylonia), he has a vision of a written scroll which he eats (Ezek. 2:8 to 3:3): he has detailed visions of what is taking place in Jerusalem, a thousand miles away (Ezek. 8:3): he sees angels marking the foreheads of those who are to be preserved during the Sack of Jerusalem, and slaughtering those who are to die (Ezek. 9:1-7). As this is well before the chronological date of the actual fall of Jerusalem, this is obviously a vision of some kind. He sees a valley full of dry bones (Ezek. 37:1-2): he assures us that this experience was when 'the hand of the Lord was upon me', and that he was brought out 'by the Spirit of the Lord', so that we know he means us to understand it as a vision. Therefore we are not surprised when he sees bone coming to bone, and flesh and sinew covering them, and finally sees life returning. Ezekiel is an imaginative poet of the first order, as well as a prophet with a most unusual psychological make-up, given to visionary experiences foreign to most of us: he is not a sober observer and recorder of scientific fact, and it is unfair to his aims to make him so. It is, of course, just as unfair to the author of Genesis to mistake his poetical description of creation in Genesis, chapter 1, written to convey religious truth, for a scientific account of the origin of the earth, with event following event in the correct scientific order (Däniken, page 51). This is as though we were to estimate the beauty of Niagara Falls by its capacity to supply water for a cup of tea.

Far From the Madding Crowd

But there is another good reason (even if all the above had not been so) why we can be sure that Ezekiel is not describing an actual space-craft. He distinctly says that this incident (not to say 'vision') occurred in the midst of his people, refugees settled at Tel-Abib on the irrigation canal Chebar (Ezek. 1:1). In the middle of a refugee settlement, packed on a 'Tell-Abubi' (to use the Mesopotamiam form of the word), a mound left by a storm on the banks of a river, is it conceivable that one man could have seen a space-craft landing, without all his fellows likewise observing the phenomenon? Yet it seems to have been a private experience, shared by nobody else: this would suit a vision, not an external event. Further, Babylon was densely inhabited at the time, especially along the river banks; the Babylonians were heirs to one of the greatest astronomic traditions of the ancient world; and Nebuchadnezzar was still at the height of his power, with the Neo-Babylonian Empire still unchallenged. Is it conceivable that, with their ceaseless scanning of the sky for astrological portents, no Babylonian astronomer would have observed the coming of the space-craft? Or that, if he had, such a unique event would have gone unrecorded? If it was a physical event, and as noisy and bright as Ezekiel suggests, it would have been visible and audible over a large area of Babylon: and it apparently was not. No, regretfully, we must give up all idea of a space-ship, and return to what Ezekiel himself said it was — a very moving and highly symbolic vision of God, in all His power, mobility, wisdom, and omnipresence. Of course, illustrations of the chariot taken from medieaval Bibles, purporting to show more clearly what Ezekiel saw, and used in the film as clear evidence that Ezekiel saw an actual space-craft, are worthless as evidence. Not only are they two millenia later than Ezekiel: they are only an imaginative artist's interpretation of the same Bible text, the same primary evidence, as lies before us. We might venture to say that they are only one imaginative artist's interpretation of another imaginative artist's interpretation, and thus two removes from reality itself. One might as well study Michelangelo's famous statue of David with hopes of making new historical discoveries about the kingdom of Judah in 1000 B.C.

Was the Ark a Condenser?

A second major area of von Daniken's argument where he uses Biblical evidence is in connection with the Ark, the small chest made of acacia wood and covered with gold, in which were, by tradition, deposited some of Israel's most holy objects (the two small tablets engraved with laws, for instance). Von Däniken introduces this on page 58 of his book: his argument is that the Ark was actually a condenser, highly charged electrically, which therefore necessitated protective clothing and careful handling. If these precautions were ignored, says von Däniken, death by electric shock could and did follow. Further, he sees the two golden cherubim, on the lid of the Ark, as some kind of loudspeaker or transmitter-and-receiver, by which Moses kept in touch with his particular 'spacemen'. I shall leave scientists, better qualified than I, to show the unlikely nature of this sort of condenser, or of any danger arising from it: I shall simply deal with some very obvious gaps and mis-statements in the evidence produced from scripture.

In the first place, Exodus never once says that the Ark was surrounded by flashing sparks, nor that Moses used it (or the cherubs) as a transmitter; nor does it mention any alloys whatsoever as used in the construction; nor does it detail any special protective clothing or footwear to be worn by those transporting it. These are four plain mis-statements, all in pages 59-9, that could have been corrected by simple reference to the English text of Exodus; no knowledge of Hebrew, or consultation of abstruse records is required here. It is true that there are detailed instructions for the making of the Ark in Exodus 37:1-9, but these show clearly no two gold plates, as postulated by von Däniken, but a simple box of acacia wood, covered with golf leaf. It is to be presumed that the author is really thinking of the two tablets of the law, deposited, by tradition, within the Ark (Deut. 10:5); but these are of stone, not of gold (Ex. 34:1).

It is also true that elsewhere in the book we have detailed descriptions of complicated priestly dress (Ex. 39:1-31, but there is no reference whatsoever to this being worn when engaged in physical labour like carrying the Ark: such robes were for great occasions like sacrifice. There is, of course, no evidence that such clothing was in any way protective or insulatory; and there is actually no reference to shoes at all. Again, while it is correct that Uzzah fell dead after touching the Ark (2 Samuel 8:2) it is far more likely that his death was due to superstitious terror at the realisation of what he had done in breaking a strict tabu, rather than due to electrical impact. Finally, while we are told that God had promised to meet with Moses before the Ark, and to speak with him from above the cherubim (Ex. 25:22) this was because it was the holiest place of all, the centre of worship, seen as the throne of the invisible God, not because of any hypothetical loud-speaker concealed there. Exactly the same phrase is used in Exodus 29:42 of God speaking to Moses at the door of the meeting-tent, where there were no possible cherubim to act as loud-speakers.

As to the story of the model of the Ark made in the States, and later destroyed because it was so dangerous, since no date and documentation is given, we may dismiss it as evidence. It sounds inherently unlikely: and, even if it were true, it would only tell us something about a modern, manufactured object, not about an object of the time of Moses.

Sons of God and Daughters of Men

Chapter 4 of *Chariots of the Gods?* contains numerous minor 'proofs', also drawn from the Bible. Admittedly, many are obscurities, if not puzzles: but more convincing solutions can be suggested in every case. For instance, the pronoun 'we' not 'I' is used of God in Genesis 1:26 probably because the Hebrew word for 'God' is a grammatical plural. There may also be a thought of the so-called 'royal We', the plural of authority. There may even be the thought that God is taking counsel with a court of angels. Just as any earthly king had his council of nobles, so in the ancient world the divine king was pictured as sitting among a group of inferior spiritual beings who formed his court: the same idea is present in Job 1:6.

The mysterious 'sons of God', mentioned in the obscure passage in Genesis 6:1-2, may also refer to this concept. In all fairness, let us admit that later Jewish exegesis saw this as a reference to fallen angels having intercourse with mortal women. But, if this interpretation is correct, then the passage must be an old piece of symbolic mythology, not to be understood literally, and so many modern commentators would understand it. It may be intended to give the impression of general lawlessness in antediluvian days, possibly even to failure to keep 'caste' in society. However, there is no guarantee that this interpretation is correct: some understand 'sons of Gods' and 'daughters of men' as referring essentially to the same people, and see the passage as simply referring to marriage and home life. Others see the two groups as two differentiated groups of humans — perhaps the 'good' line, the presumed descendants of Seth, and the 'bad' line, the presumed descendants of Cain. Others see dim memories of the mating of Homo sapiens with Neanderthal Woman in early days. Enough has been said to show that there are countless possible explanations.

Incidentally, the Bible does not say (p. 52) that these 'sons of God' were giants. It does say that in these far-off days there were 'giants' and men of 'renown' and 'mighty men' (Gen. 6:4): this was an age of heroes. This introduces the question of giants: the fact that 'giants' keep on cropping up in story in all parts of the world does not (p. 52) prove in itself that they ever existed. Fairies also crop up everywhere in folk tales; by the same argument, they should exist. It is very common in folk history to ascribe any surviving ruins, too great or too complex for the present generation to build, to the work of assumed 'giants' in the past. European literature, for instance, is full of such stories, yet we have no archaeological evidence for people of gigantic stature anywhere in Europe from the earliest times.

Von Däniken does not realise that he is falling into a similar error himself when he postulates such giants in early days to explain huge monoliths — or even assumes that his primaeval spacemen were of gigantic stature themselves. To do this is simply to bypass the need for finding a scientific explanation. However, that there were 'giants' (in a limited sense) in Palestine in historical times seems indisputable: but there is not the slightest hint that they were in any way connected with spacemen, or that they sprang from intermarriage with spacemen. Israel appears to have been a short and stocky race, at least on their emergence from Egypt: certainly, the superior stature of one tribal group of Canaan, usually called 'the Anakim' or 'the sons of Anak', made a great impression on them (Deut. 9:2). Just so, the Mediterranean men of the late Roman Empire reacted to the great shambling Goths who poured down from the north: but in neither case is there anything superhuman or mysterious about the tall strangers. Later in Israel's history, a close-knit military group (the Philistines) seem to have carried on this tradition of size: mention of Philistine 'giants' is frequent (witness Goliath, David's opponent, 1 Sam. 17:4) but again there is no hint of any supra-terrestrial origin for them. The average height of the Watutse people in Rwanda today, or the Dinka people in the Sudan, is well over six feet: mere size proves nothing as to non-terrestrial origins.

Much more could be said: but enough has already been outlined to show that the so-called 'proofs' from the text of Scripture are groundless. At best, they are a series of questions, pointing to problems or uncertainties: at worst, they are misconceptions, based on a failure to read the actual text carefully.

It Does Not Compute

PHILLIP GROUSE

P. J. Grouse, M.Sc., B.D., is Acting Master and Dean of students at New College, Senior Lecturer in Computer Studies at the University of New South Wales.

Shades of the film '2001' with its visual imagery combined with science fact/fiction, our prehistoric past and alien monoliths!

I am asked to comment, as a scientist, on what purports to be a work of science. *Chariots of the Gods?* abounds with the in-words of science. Von Däniken speaks with reverence for 'hypothesis', 'inference' and 'theory'. Facts, formulae and figures abound; but I am at a loss to see it as a serious scientific exercise.

A potentially viable scientific hypothesis should be capable of standing upon its own feet, relying upon the observed data and carefully formulated, logically derived inferences which themselves are capable of independent analysis. Clearly such a process is in the mind of the author, however his inferences appear without the benefit of the accepted due processes of logical derivation. Such derivations that do exist lie shrouded behind a cloud of emotive assertion.

In approaching a work such as this in an 'impartial' yet critical vein, it is so easy to fall into the same traps as those into which the author himself has fallen. Our natural reaction to an emotional outburst is to respond in kind. In this sense science is 'unemotional'. It cannot afford that luxury. Nor can it afford to accept as unquestioned the 'data' presented as the basis for some elegant superstructure.

Furthermore, a critic in this area must be equally impartial in his criticism of the other critics! I fully concur with the comments made by Garner[1] in his brief analysis of the book. Indeed, as a Christian I share his conviction that such an explanation of what we know as Christianity is absurd.[2] Nevertheless criticism itself must be precise, moreso than its target. Thus von Däniken asserts[3] that if one divides the area of the base of the pyramid of Cheops by twice its height one arrives at the constant π (= 3.14159). Garner[4] questions this assertion by showing that 570,780 square feet divided by 962 feet is 593.3, a value far removed from 3.14159. And not surprising, since the value derived depends entirely upon the selected unit of length. The foot was not a standard unit of measurement in Cheops' day, however one still fails to get π even if the cubit is used. The required unit must have a length of 128.77 cubits to achieve the desired result. I am unaware of any such standard. It would be more significant if the division of two linear measurements of the pyramid produced the required constant, but even if such were the case it would hardly be evidence for extraterrestrial interference.

For all his evidence, von Däniken gives no direct instance of an 'artifact from space'. Copies there may be, but where are the time capsules, the remnants of some electronic or electromagnetic technology? The history of a race may be found in its garbage! Future generations will be overwhelmed with the wealth of data from our own garbage cans. The glass, the quartz and silicon will last for millenia bearing the story of our technology. If von Däniken could produce one authentic advanced technological artifact from the past then there is a serious case for further research.

Although I am a Christian, it is not on Christian but on purely scientific grounds that I am sceptical of von Däniken's claims. Indeed I would personally be quite excited if some genuine find were made pointing to the appearance on earth of beings from another planet. While such a discovery would carry no present theological implication its scientific importance would be exceptional.[5]

Admittedly the process of translation has done little to enhance the work,[6] but mathematics requires very little in the way of translation. Thus the simple computation of the number of miles in a light year is fouled up[7] by a factor of 60. His computation of the probable number of inhabited planets in this universe is compounded guesswork, ignoring the simple fact that terrestrial intelligent life has existed for a mere 0.0001% of the life of this planet. This figure shrinks to 0.000001% if we limit ourselves to technological intelligent life. Furthermore, 'inhabited' need not imply a capacity for space travel.

A marvellous piece of mathematical one-upmanship enables the reader to compute the velocity of a relativistic rocket given the jet speed, the velocity of light and the fuel load at launching.[8] When such formulae are presented in normal scientific literature they are offered as the conclusion of a mathematical justification and/or as the basis for specific calculations or further inferences. Although no doubt valid,[9] the formula stands in glorious isolation, neither justified nor relevant.

On the same page he computes (by means of the Lorentz formula) that travellers on a space ship moving at 99% of the velocity of light would experience only 14.1 years while 100 years pass on earth. The figure is correct, but the effects of acceleration and deceleration are completely ignored, particularly embarrassing if you have arrived at your destination with empty tanks and you are moving at 184,000 miles per second.[10]

The technical and computational abilities of earlier civilisations are so frequently called into doubt that we, like von Däniken, are surprised to find that the principles of mathematics have been known for millenia. The abacus, a manually operated digital calculating device with extreme precision (e.g., 15 digits) was in all probability used by the Babylonians as early as 2200 B.C. Herodotus records its use in Egypt around 460 B.C. It was probably introduced to China via India around the 12th century A.D.[11]

The analogue computer discovered off the Greek island of Antikythera in 1901[12] and dated about 82 B.C. is of much later origin and of lower precision when compared with the abacus. The importance of the Antikythera computer lies in the light which it sheds on the mechanical and astronomical understanding and ability of the Greeks. There is no need to invoke 'gods' to explain their insights.

Space forbids the detailed examination of each 'datum' presented as evidence by von Däniken. In addition, detailed study of a number of assertions is rendered difficult by the absence of references. It is disconcerting to find such an omission in a work which purports to advance a major scientific hypothesis. Nevertheless I will conclude my comments with a few observations on some matters which inspired me, on the first reading of the book, to scribble rapidly in the margin!

To date, no evidence has been offered to support the assertion (in the context of meteorites) that bacteria millions of years old reawaken to new life[13] nor that spores float through the universe awaiting capture by a planet's field of gravitation. Such a discovery would be of major importance. The detection of certain amino acids in a number of meteorites[14] does not imply the prior presence of some carbon-based life form. While these trace chemicals are relatively simple portions of terrestrial organic matter, they may be produced by a number of non-biological means. It is a long step from amino acids to the life giving chains of DNA and RNA.

Scientists have never known that tachyons (particles that travel faster than light) *must* exist.[15] The assertion has been that they *may* exist. Only the detection of such a particle could change the *may* into a *must*.

The Venusian year is not 584 days[16] but 224.7 days. The Venusian *synodic* year (its period with respect to the earth — not the Sun) is indeed 584 days and can be determined by simple observation. No astronomical insights are needed. The Mayan Venusian formula[17] is therefore in keeping with what is already known about the Mayan civilization.

Finally, the assertion that Phobos is hollow[18] is now not taken seriously as a result of the physical irregularity of the object as revealed by Mariner 9.[19]

REFERENCES

1. GARNER, G. G., *Chariots of the Gods?—A Critical Review,* Aust. Inst. of Archaeology, 1972.
2. Perhaps in one sense the Christian concept and doctrine of the incarnation of Christ could be regarded as an 'extra-terrestrial visit' from 'outer space', but let it be clearly understood that this 'visit' was planned from the beginning of time, that its medium was not a space capsule but a human uterus, that the 'visitor' was not an alien but one of us, and that he came not to give us a 'helping hand' but to die in our stead.
3. VON DANIKEN, p. 99.
4. GARNER, p. 10.
5. In passing, it would seem reasonable that the Lunar surface holds more promise than the Earth when we consider the comparatively long periods that an object may remain undisturbed by an environment devoid of water, winds or

vandals. The Moon is, furthermore, an ideal Earth observation post. Any artifact discovered there or here would scarcely 'shake the foundations of our religions' as von Daniken suggests (p. 71). It is worth noting the comments of Wernher von Braun quoted on pp. 175 ff. Perhaps von Daniken is unaware that von Braun is a Christian believer!

6. e.g., the words 'Utopia' and 'Utopian' never seem to be relevant in their contexts. Again: 'Our Earth is the only *star* with human beings', etc..
7. Von Däniken, p. 15.
8. Ibid, p. 23.
9. *t* refers to the *mass-ratio* not *fuel load at launching.*
10. VON DANIKEN, p. 23.
11. CRAWFORD, F. R., *Introduction to Data Processing,* N. J., Prentice-Hall, 1968, p. 15.
12. COLE, R. Wade, *Introduction to Computing,* N.Y., McGraw-Hill, 1969, p. 10 *and* VON DANIKEN, p. 128.
13. VON DANIKEN, p. 19.
14. LAWLESS, J. G., *et al., Organic Matter in Meterorites,* Scientific American, June, 1972, p. 38.
15. VON DANIKEN, p. 141.
16. Ibid, p. 75.
17. Ibid., p. 76.
18. Ibid., p. 154.
19. *Three Spacecraft Study the Red Planet,* article in *Sky and Telescope,* January, 1972, p. 16.

Was Santa a Spaceman?

ANONYMOUS

The contributor of this paper asks to be allowed to be anonymous out of — he says — diffidence. He was a used-car salesman for many years but has recently devoted himself to full-time studies in esoteric anthropology and occult archaeology. He describes himself as a very, very distant friend of Erich von Däniken's.

Mr. von Däniken believes, deeply, that in prehistoric times, and in several places, the human race was visited by extra-terrestrial strangers. These strangers, whose technology and morality were more highly developed than primitive man's, interbred with selected earthlings, and transmitted to their hybrid offspring the more accessible elements of their alien culture. This was the origin of *Homo sapiens* as distinct from mere *Homo,* whom they treated with some harshness, especially in places like Sodom and Gomorrah.

Subsequently, Mr. von Däniken submits, the strangers have revisited planet earth to see how things were going — though with less frequency latterly. Unhappily, things have not gone all that well, for men have translated the spacemen into gods, their technology into magic, and their morality into religion. Possibly (though Mr. von Däniken does not explicitly volunteer this) it is disappointment at their failure which has limited their more recent visitations to very rapid surveys from flying saucers.

In presenting his case, my friend Erich had neither time nor space to marshal all the evidence he had available. He chose to concentrate on the more striking archaeological mysteries, and had to put aside the many thousands of homely facts and phenomena which might also have served his argument.

In a way, that was a pity. Erich, although he frequently warned his readers against conservative pre-historians and anthropologists, has left himself vulnerable to their counter-attack. As the other contributions to this present book so clearly show, his own innocent presentation of the clear facts was not proof against the confusion and complications which alleged 'experts' were able to introduce by eliciting other facts and sharper logic.

I am, therefore, moved to come to his rescue by drawing attention to some more evidence in his favour, and that of a homely and familiar kind. It is not in ancient Ephesus or in distant Kuyundjik that I find my further argument, but at home, in the hearts and minds of the common people of the western world . . . which, for reasons not shown on the Piri Re'is map, includes Australia.

I would like to mention, as a preliminary, some of the concepts which we use in our own daily lives. Why, in a law court, is a finding *handed down?* Could it not be because the hatch of a space-vehicle is always several feet above the ground? Why do temperatures go *up* when things become hotter, and not *down*? What suggested the imperfectly spherical shape of the traditional goldfish bowl, if not the top part of a space-suit?

Again, what is a church spire if it is not an invocatory replica of an interstellar rocket? Who or what suggested the dimension of a cricket-pitch and the rules of the game played on it which involves hitting *a spherical red object* towards the outer limits? Who, indeed!

These, I say, are preliminary questions. It is not intended to investigate them all here. (Indeed it is barely necessary to, since they imply their own answers.) What I would like to do particularly, in support of my friend Erich, is to make a brief but suggestive study of the figure of Santa Claus in our culture.

Santa Claus, as far as can be discovered, has never been thoroughly investigated by orthodox antiquarians, anthropologists and archaeologists, obsessed as they are by their own vested interests in what they call the truth. Still less has this venerable figure been subjected to the concentrated attention, which one day *must* come, of spectrographic analysts, geophysicists and volcanologists. Only the eccentric anthropologist Claude levi-Strauss comes close to mentioning the possible importance of Santa Claus to human development.[1]

And yet, as soon as the most obvious elements of his legend are mentioned — those elements which are known even to the smallest children in our civilization — it becomes clear that 'Santa Claus' is nothing else than a folk memory of just such an inter-stellar visitor as Erich has described.

He is, to begin with, a technological wonder-worker. Popular account endows him with a workshop in which he, with the help of 'gnomes' (surely cybernetic devices of some kind) manufactures children's toys in incredible quantities. By the same popular account, his workshop is stationed at the North Pole. Considering that the North Pole is 'up' on a map, and that the North Pole was to earlier Man the ultimate in remoteness, it is not the least fanciful to see 'the North Pole' as *really* being some sort of orbiting space vehicle.

A typical drawing of Santa Claus from the wall of an Australian kindergarten. What are the curious lines on the body? What are the projections from the 'head'? Notice, too, how large and dominating the figure is made to be.

As to the way Santa Claus is said to travel, it is plain that, in spite of the explanations given by the enemies of truth, we have a naive description of some sort of rocket. Santa's vehicle travels at unbelievable speed. Even if some element of fantasy is allowed, and the number of his delivery stops is reduced, he must still be able to make a complete circuit of the earth, visiting every major population centre, within the space of one night.[2] The only workable explanation of the legend is that there was once a visiting space man who regularly did this.

And not only is his speed significant. The vehicle which he is said to use is *wheelless,* and propelled by six exotic animals, one of which has a red nose, and all of which breathe a misty smoke. A small rocket propelled by a photon accelerator would give simple people just that impression!

Santa Claus (or Father Christmas, or Chris Cringle, for he has as many names as the various 'gods' who appeared to the people of the Nile Valley) makes his landings from his space-vehicle by way of chimneys. It does not take much imagination to realise that a chimney, except for its size and complexity and the material from which it is made, is *exactly* the same as the vast gantries which are nowadays used to support rockets at Cape Kennedy. Is it not easy to guess that primitive people, seeing such a gantry, would imagine it to be a chimney of some kind?

By a very ancient tradition, Santa begins his annual travelling at midnight, G.M.T., on December 24th. It so happens that that time and date correspond exactly to the Winter solstice in the latitude and longitude of Isfahan in the year 11,476 B.C. So we are able to give the exact time and date at which the original space-Santa made his first visit to Earth.

The dress which Santa is supposed to wear is not unlike the dress in which a space-traveller would have had to be clad. He wears red garments lined with some fur-like material, held in at the middle by a broad belt fastened by a huge shining buckle-like object. Over his head he wears a red helmet which ends in a mysterious small white ball. His face is covered by a mask, which the ignorant have recorded as being like a very thick, white beard.

This is indeed a very mysterious stranger! He cannot be explained away as anything but a visitor from a distant place in space.

But the most astonishing thing of all about him is the way — the exact way — in which the importance he has in a country's culture corresponds to the degree of development of its technology. The Australian Aranda have no Santa Claus figure: they are the most primitive people on earth. The United States, the country richest in the world for versions of the Santa Claus legend, has the most advanced technology. Japan, which a century or so ago had neither Santa Claus nor technology, has shown a fantastic development, and now leads the world in the production of micro-electronic equipment *and* of plastic figures of Santa Claus.

It is time to open our hearts and minds, and to realise that Erich von Däniken is right! Everywhere there is evidence that space-travellers have visited earth and given us their knowledge, if only we will see it.

REFERENCE

1. In an unwritten chapter of *Les Structures elementaires de la parente.* (Paris, 1949).
2. Like Shakespeare's Ariel (= antenna?) who could 'put a girdle round the earth in sixty minutes'.

Clip and Mail This Special Shipping Label and...

SHIP TO:

PRINT YOUR NAME HERE

YOUR ADDRESS

CITY STATE ZIP #

CONTENTS: 6 INDISPENSABLE PAPERBACKS

We will send you all six of these widely-acclaimed reference paperbacks for only $5.45

A $27.30 HARD-COVER VALUE

PERFECT FOR HOME, SCHOOL OR BUSINESS OFFICE!

SEE OTHER SIDE ➧